FORESTRY COMMISSION BULLETIN 92

# Poplars for Wood Production and Amenity

*J. Jobling*
*Silviculturist, Forestry Commission*

LONDON: HMSO

© *Crown copyright 1990*
   *First published 1990*

ISBN 0 11 710285 7
ODC 83: 273: 176.1 Populus: (410)

KEYWORDS: Amenity,
Broadleaves, Forestry

Enquiries relating to this publication
should be addressed to:
The Technical Publications Officer,
Forestry Commission, Forest Research Station,
Alice Holt Lodge, Wrecclesham,
Farnham, Surrey GU10 4LH.

**Front Cover:** *Populus* 'Robusta'. (*38973*)

# Contents

| Chapter | Page |
|---|---|
| **Summary** | v |
| **Résumé** | vi |
| **Zusammenfassung** | vii |
| **Preface** | viii |
| **1. Botanical classification of the genus *Populus* and species' distribution** | 1 |
| Botanical classification | 1 |
| Characteristics and distribution of the sections of the genus *Populus* | 1 |
| Species' distribution | 1 |
| **2. Description of species and cultivars** | 4 |
| Section *Turanga* | 5 |
| Section *Leuce*, subsection *Trepidae* – aspens | 5 |
| Section *Leuce*, subsection *Albidae* – white poplars | 7 |
| Hybrids between subsections *Trepidae* and *Albidae* | 8 |
| Section *Aigeiros* – black poplars | 9 |
| Section *Tacamahaca* – balsam poplars | 19 |
| Hybrids between sections *Tacamahaca* and *Leuce* | 26 |
| Hybrids between sections *Tacamahaca* and *Aigeiros* | 26 |
| Section *Leucoides* | 31 |
| **3. Field recognition of the chief poplars grown in Britain** | 33 |
| Botanical key | 33 |
| Sex and date of leafing | 36 |
| **4. Choice of site** | 38 |
| Climate | 38 |
| Soil and soil moisture | 39 |
| Situation | 40 |
| **5. Plant production** | 44 |
| Hardwood cuttings | 44 |
| Root cuttings | 46 |
| Root suckers | 47 |
| Softwood cuttings | 47 |

|  |  |
|---|---|
| Seed | 48 |
| The nursery | 49 |

## 6. Planting and establishment    50
Planting    50
Weed control    51
Fertilising    52

## 7. Spacing, thinning and pruning    54
Spacing    54
Thinning    56
Pruning    56

## 8. Rate of growth and yield    58
Rate of growth    58
Yield    59

## 9. Poplars and farming    63
Line planting    63
Arable cropping    63
Grazing    65

## 10. Characteristics, properties and uses of poplar wood    66
Wood characteristics    66
Wood properties    66
Uses of wood    67

## 11. Pests and diseases    70
Insect pests    70
Diseases    71
Wind damage    72
Squirrel damage    73

**References**    74

**Appendix I** The International Poplar Commission    76

**Appendix II** New poplar clones from Belgium    76

**Index**
Subjects and technical terms    77
Species and cultivars    79

# Poplars for Wood Production and Amenity

## Summary

The eleven chapters comprising this Bulletin cover the botany, cultivation, performances and utilisation of poplars and poplar timber. The genus *Populus* comprises some 32 species classified according to their botanical characters into five sections and one sub-section. Of these, the sections *Aigeiros* (the black poplars) and *Tacamahaca* (the balsam poplars) are of commercial significance to poplar growers in Britain and Europe. The species are described in Chapter 2 together with all the hybrids and cultivars thought to have been or to be noteworthy in Britain. The species are described first within their respective sections followed by their specific cultivars and then by the hybrids within and between sections. In Chapter 3 a key attempts to assist the field recognition of most of the commercially important poplars grown in Britain based on morphological characters and crown form.

In Chapter 4 the choice of sites suitable for poplar cultivation is explained, with advice on the field recognition of both suitable and unsuitable sites. Plant production is covered in Chapter 5 which describes the main methods and vegetative and seedling reproduction and nursery practice. The next two chapters cover the silviculture of poplars including establishment and tending practices in Britain and Europe and, in Chapter 7, recommendations on spacing, thinning and pruning to meet various objectives.

Rates of growth of poplars under various conditions are discussed in Chapter 8 with examples of some outstanding cultivars and with summaries of research data currently available on yields for biomass and timber. In Chapter 9 the use of poplars in association with farming is described with particular reference to the development of agroforestry practice in Britain during the 1960s and 70s. The characteristics and properties of poplar timber and its uses are then described.

The final chapter provides a general description of the more important insect pests, fungal and bacterial diseases of poplar. The Bulletin concludes with a short reference list, appendices and indexes.

# Peupliers pour la Production de bois et l'Agrément

## *Résumé*

Les onze chapitres composant ce Bulletin portent sur la botanique, la culture, les performances et l'utilisation des peupliers et du bois de peuplier. Le genre *Populus* comprend 32 espèces classées selon leurs caractères botaniques en cinq sections et une sous-section. Parmi celles-ci, les sections *Aigeiros* (les peupliers noirs) et *Tacamahaca* (les peupliers baumiers) sont importantes sur le plan commercial pour les producteurs de peupliers en Grand-Bretagne et en Europe. Les espèces sont décrites au Chapitre 2 ainsi que tous les hybrides et les cultivars qui ont été ou qui sont dignes d'attention en Grande-Bretagne. Tout d'abord, les espèces sont décrites au sein de leurs sections respectives puis selon leurs cultivars spécifiques et enfin selon les hybrides obtenus dans une même section ou en mêlant deux sections différentes. Le Chapitre 3 montre l'effort entrepris pour favoriser la reconnaissance, sur le terrain, des peupliers importants pour le commerce et produits en Grande Bretagne, selon des critères morphologiques et la forme de leur cime.

Le Chapitre 4 explique le choix des sites propres à la culture des peupliers et donne des conseils pour reconnaître sur le terrain les sites appropriés et inappropriés. La production de la plante est abordée au Chapitre 5 dont le but est de décrire les méthodes principales et la reproduction végétative, et celle des jeunes plantes ainsi que l'utilisation des pépinières. Les deux chapitres suivants concernent la sylviculture des peupliers, y compris les pratiques d'établissement et d'entretien en Grand-Bretagne et en Europe; et, au Chapitre 7, des conseils sont donnés en ce qui concerne l'espacement, l'éclaircissement et l'élagage afin de réaliser certains objectifs.

Les taux de croissance des peupliers dans certaines conditions sont abordés au Chapitre 8 et illustrés par des exemples de cultivars remarquables et des comptes-rendus de recherches scientifiques, actuellement disponibles, sur la production de biomasse et de bois d'oeuvre. Le Chapitre 9 décrit l'utilisation des peupliers en rapport avec l'agriculture, en particulier en se référant au développement des activités d'agro-sylviculture en Grande-Bretagne, durant les années 1960 et 1970. Les caractéristiques du bois de peuplier et son utilisation y sont expliqués.

Le dernier chapitre fournit une description générale des insectes nuisibles les plus importants, ainsi que des maladies bactériennes et fongiques du peuplier. Le Bulletin s'achève par une courte liste de références, des annexes et des tables alphabétiques.

# Pappeln für die Holzproduktion und Landschaftserhaltung

## *Zusammenfassung*

Die elf Kapitel, die dieses Bulletin enthält, befassen sich mit der Botanik, dem Anbau, der Leistung und dem Gebrauch von Pappeln und Pappelholz. Die Gattung *Populus* unfaßt etwa 32 Arten, die je nach ihren botanischen Merkmalen in fünf Sektionen und eine Sub-Sektion eingeteilt werden. Von diesen sind die Sektionen *Aigeiros* (Schwarzpappel) und *Tacamahaca* (Balsampappel) für die Papelanbauer in Großbritannien und dem übrigen Europa von kommerzieller Bedeutung. In Kapitel 2 werden die Arten beschrieben, einschließlich all der Hybriden und Sorten, die im Zusammenhang mit Großbritannien als erwähnenswert erachtet werden. Die Arten werden zunächst innerhalb ihrer jeweiligen Sektionen beschrieben, gefolgt von ihren speziellen Sorten und dann von ihren Hybriden innerhalb und zwischen den einzelnen Sektionen. Der in Kapitel 3 enthaltene Schlüssel versucht, durch Beschreibung der morphologischen Merkmale und Kronenformen, bei der Bestimmung der meisten in Großbritannien wachsenden Pappeln von kommerzieller Bedeutung behilflich zu sein.

In Kapitel 4 wird die Wahl der Standorte, die für den Anbau von Pappeln geeignet sind, erklärt und es werden Ratschläge für das Erkennen von geeigneten und ungeeigneten Standorten gegeben. Die Pflanzenproduktion wird in Kapitel 5 behandelt, in dem die wichtigsten Methoden, die vegetative und die generative Vermehrung sowie Baumschultechniken beschrieben werden. Die folgenden zwei Kapitel beschäftigen sich mit dem Waldbau von Pappeln in Großbritannien und dem übrigen Europa, einschließlich Anbau und Pflege, und Kapitel 7 enthält Empfehlungen für die räumlichen Abstände, die Durchforstung und das Beschneiden, um verschiedene Ziele zu erreichen.

In Kapitel 8 werden die Wachstumsraten der Pappeln unter verschiedenen Bedingungen besprochen; außerdem enthält es Beispiele für einige hervorragende Sorten und eine Zusammenfassung der gegenwärtig zur Verfügung stehenden Forschungsdaten in bezug auf Biomassen- und Holzerträge. In Kapitel 9 wird die Verwendung von Pappeln in Verbindung mit der Landwirtschaft beschrieben unter besonderer Berück-sichtigung der Entwicklung der agrarforstwirtschaftlichen Methoden in Großbritannien während der sechziger und siebziger Jahre. Anschließend werden die Eigenschaften und Merkmale des Pappelholzes sowie seine Verwendungsmöglichkeiten beschrieben.

Das letzte Kapitel enthält eine allgemeine Beschreibung der wichtigeren Arten des Schädlingsbefalls sowie der bei Pappeln auftretenden Pilz- und bakteriellen Krankheiten. Das Bulletin schließt mit einer kurzen Bibliographie, Appendizes und Registern.

# Preface

It is with sadness that we publish this Bulletin after the untimely death of John Jobling in October 1987. For more than three decades John was the Forestry Commission's expert on poplars and their cultivation and this work largely represents a distillation of the knowledge he gained. It is very much an updating and expansion of Forestry Commission Leaflet 27 *Poplar cultivation*, which John wrote, and provides a suitable successor to Tom Peace's Forestry Commission Bulletin 19 *Poplars* published nearly 40 years ago (in 1952).

The Bulletin was not complete when John died and we are grateful to Mr Arnold Beaton for further work on Chapter 2, for writing Chapter 9 about poplars and farm forestry, and for making many helpful suggestions to the text. Dr Julian Evans drafted Chapter 7 and was responsible for bringing the text to its final form. Staff of Pathology and Entomology Branches drafted Chapter 11.

The Bulletin follows closely John Jobling's original structure and we trust it will serve the needs of poplar growers, whether for timber, farm or simply pleasure, for many years.

*John Jobling*
*1924 – 1987*

# Chapter 1
# Botanical Classification of the Genus *Populus* and Species' Distribution

## *Botanical classification*

The genus *Populus*, the poplars, and the genus *Salix*, the willows, together form the family *Salicaceae*. The *Salicaceae* are dioecious woody plants, with the wholly male or the wholly female flowers produced in catkins on separate plants. The male catkins are usually more closely packed than the female and they have yellow, bright red or purple anthers. Female catkins are green and lengthen until mature, when the seeds emerge. The seeds are small and surrounded by long, cottony-like filaments which aid wind dissemination.

Poplars and willows differ in several respects. Their main distinguishing features are summarised in Table 1.1.

## *Characteristics and distribution of the sections of the genus* Populus

The genus *Populus* is divided into five sections. They differ enormously in distribution, and in cultural and economic importance. Their characteristics and distribution are listed in Table 1.2.

## *Species' distribution*

The genus *Populus* contains about 30 species. They occur throughout the northern hemisphere in most cold and temperate zones between the subarctic and sub-tropical regions. Though a few species, in central and eastern Asia, are not well known, most have been well documented in technical literature. Several species, in North

**Table 1.1.** Characteristics of the genera *Populus* and *Salix*

|  | *Populus* | *Salix* |
|---|---|---|
| *Leaves* | Ovate, orbicular, rhomboid, triangular (deltoid), sometimes lanceolate. Great variation in size and shape on same plant (foliar polymorphism). Petiole long. | Long; ovate-lanceolate. Shape uniform. Petiole short. |
| *Buds* | Several outer scales. Terminal bud always present and larger than lateral buds. | One outer scale. Terminal bud often absent or little developed. |
| *Shoots* | Cross section circular or angular. Pith cross section five sided (pentagonal). | Cross section circular. Pith cross-section circular. |
| *Flowers* | Before leaves. Perianth oblique cup-shaped disc without nectaries. | Before or after leaves. Perianth usually absent, one or two small nectaries. |
|  | Bracts toothed or laciniate, soon fall. | Bracts entire, do not fall. |
|  | Stamens numerous, 4 to many (50), red anthers. | Stamens few, 2 to 12, yellow anthers. |
|  | Catkins pendulous. Wind pollinated. | Catkins usually erect. Insect pollinated. |

**Table 1.2.** Characteristics and geographical distribution of the sections of the genus *Populus*

| Section and Sub-section | Leaves | | | Buds | Flowers | | Distribution |
| --- | --- | --- | --- | --- | --- | --- | --- |
| | Long shoots | Short shoots | Petioles | | Male | Female | |
| **Turanga** | Lanceolate-linear; Lanceolate-deltoid-rhomboid | Rounded, broad | Round, flat close to leaf blade | Small, hairy | Red catkins, Stamens 15-25 | Stigmas 3 | Central Asia, North Africa |
| | without translucent margins | | | | | | |
| **Leuce** Trepidae | Oval, slightly hairy, greyish | Rounded, strongly toothed, smooth | Strongly flattened | Small, appressed, dry | Catkins 4-10 cm Stamens 5-20 | Catkins 5-10 cm Stigmas 2 | Northern and mountainous regions of Europe and Asia; North America |
| | without translucent margins | | | | | | |
| Albidae | Large, three lobed or palmate, dense white hairs on underside | Small, rounded, strongly toothed, white hairs on underside | Roundish | Small, covered with dense white hairs | Catkins 8-10 cm Stamens 6-20 | Catkins 2-10 cm Stigmas 4 | Central and Southern Europe, Western and Central Asia, North Africa |
| **Aigeiros** | Large, deltoid, cordate | Small, deltoid, rhomboid | Flattened | Fairly large, lightly sticky, shiny | Reddish catkins 5-8 cm Stamens 10-60 | Catkins 8-30 cm Stigmas 2-4 | Europe, Western Asia, North Africa; North America |
| | with translucent margins | | | | | | |
| **Tacamahaca** | Small-large, ovate-lanceolate, underside silvery-brownish | | Round | Large, sticky aromatic | Red catkins 5-8 cm Stamens 15-60 | Catkins 10-15 cm Stigmas 2-4 | Asia; Northern North America |
| | without translucent margins | | | | | | |
| **Leucoides** | Very large, cordate, underside violet when young | | Round | Very large, fairly sticky shiny | Catkins 5-15 cm Stamens 12-30 | Catkins 8-20 cm Stigmas 2-3 | Eastern and Central Asia; Eastern North America |

America and Europe, have a very extended range and display considerable botanical variation. In some instances sub-species and varieties have been described. The principal species in each section and their natural distribution are listed in Table 1.3.

Two species are certainly native of Britain. They are *Populus tremula* L., the aspen of Europe and Asia, and *P. nigra* L., the black poplar of Europe and Asia. A third poplar, *P.* × *canescens* Smith, grey poplar, is probably native in southern England. The best known poplars in Britain, and elsewhere in western Europe, are cultivated varieties (cultivars) propagated clonally by vegetative means.

**Table 1.3.** Geographical distribution of species of the genus *Populus*

| Section/sub-section | Species | Distribution |
|---|---|---|
| **Turanga** | euphratica | North Africa (Atlas Mountains) to Central Asia (Altai Mountains); 45°N to equator |
| **Leuce** Trepidae | adenopoda | Central and West China |
| | grandidentata | North-east North America |
| | sieboldii | Japan |
| | tremula | Europe, Western Asia and North Africa |
| | tremuloides | North America |
| Albidae | alba | Central and Southern Europe to Central Asia and West Siberia; North Africa |
| | tomentosa | North China |
| **Aigeiros** | deltoides | Eastern North America |
| | fremontii | Western United States of America |
| | nigra | Europe to Central Asia; North Africa |
| | sargentii | North America (east of Rocky Mountains from Saskatchewan to New Mexico and west Texas) |
| | wislizeni | United States of America (west Texas, New Mexico) |
| **Tacamahaca** | acuminata | United States of America (Montana and South Dakota to New Mexico and Arizona) |
| | angustifolia | Western United States of America east of Rocky Mountains |
| | balsamifera | North-east North America |
| | cathayana | North-west China to Manchuria and Korea |
| | ciliata | Himalayas |
| | koreana | Korea |
| | laurifolia | Siberia, Mongolia |
| | maximowiczii | North-east Asia, Japan |
| | purdomii | North-west China |
| | simonii | North and west central China |
| | suaveolens | East Turkey, Siberia to Russian Far East |
| | szechuanica | Western China |
| | trichocarpa | Alaska and British Columbia to California, Rocky Mountain plains (Idaho and Montana) |
| | tristis | Central Asia |
| | yunnanensis | South-west China |
| **Leucoides** | heterophylla | Eastern United States of America |
| | lasiocarpa | Central and West China |
| | violascens | China |
| | wilsonii | Central and West China |

# Chapter 2
# Description of Species and Cultivars

Since the mid-1930s, about 1000 different poplars have been raised at one time or another in nurseries in Britain in the search for clones suitable for wood production or amenity. Most of the propagation has been undertaken, on a continuing basis, by the Forestry Commission's Research Division using clones systematically introduced for disease or field trials. Occasionally other research organisations and a few commercial interests have also become involved. This sustained effort over a 50-year period can be traced directly to the Commission's early interest in the comparative behaviour of different species and hybrids. The first poplar trials were planted in 1937 and the first studies in this country on bacterial canker were started just a year or two earlier.

With only two or three exceptions all the poplars raised in Britain have been and still are obtained from sources overseas. In a few instances, well-known cultivars imported from regions where the soils and climate are similar to those in this country have been found to satisfy growers' requirements after only cursory examination. In nearly every case, however, clones have been introduced from places where growing conditions are different, in one respect or another, from those in Britain or they have been relatively new selections little known even in their country of origin. As a consequence virtually every clone has had to be repeatedly propagated and tested over several years to assess its value. Fortunately, resourceful breeding programmes have been pursued over several decades in neighbouring western European countries and rigorously tested clones, mostly hybrids bred artificially from carefully selected parents, have become increasingly available in recent years.

Inevitably, disease and winter cold damage have led to many clones being discarded. Others have been dropped on account of their excessively poor habit or slow growth. It is probable that some clones from warm countries have simply not adapted well to our comparatively cool summers. None the less, nearly 500 clones – perhaps half the number imported over five decades – are established in stool beds at Alice Holt Lodge for the provision of true-to-name herbarium specimens and reproductive material. Moreover, around 100 clones are marketed by horticulturists and poplar growers; most were acquired in the 1960s when interest in poplar culture and research reached a peak.

Poplars vary enormously in appearance and behaviour, and trees for particular uses have to be carefully selected. In Britain most poplars are planted for screening, shelter or ornament. Some of them are grown on account of their distinctive habit or leaf colour, a few because of their characteristic odour or the pleasing sound made by rustling leaves. Others are chosen thanks to their resistance to atmospheric pollution or to salt-laden winds. Some are even planted because they flush earlier than other deciduous trees. Straight stemmed, vigorous and lightly branched cultivars are much preferred for the production of merchantable wood. Most poplars are considered to be visually attractive and suitable for inclusion in landscaping schemes with other trees.

Diseases are a very important consideration in the selection of poplar species and cultivars in breeding programmes. The relative resistance of species and cultivars to the major poplar diseases is discussed in this chapter wherever appropriate and there is also a more general account of these diseases in Chapter 11.

This chapter reviews most of the species and cultivars that have been planted from time to time in Britain. Understandably, several poplars that are quite rare even in tree collections but deserve to be more widely planted have been included. In the same spirit, attention is also drawn to several new clones which, on the face of it, appear to have potential merit and should be planted here if only on a trial basis. In contrast, several poplars that were once relatively familiar but, with the passage of time, have proved to be unacceptable receive short shrift. A few clones that were cultivated for such a short space of time that they are now forgotten have been omitted altogether. Where appropriate, comment has been included on date of leafing, speed of growth, ultimate dimensions, angle of branching and crown width.

Species and their varieties and cultivars are described alphabetically by botanical section (see Table 1.3). Hybrids and hybrid cultivars are considered alphabetically after the species.

## Section Turanga

### P. euphratica Oliv., Euphrates poplar

This species, the only one in the section, is mentioned only because, without much doubt, it is the most complex poplar of all. In the first place it has an enormous geographical distribution. It occurs as far west as Morocco, extends through North Africa, through the Near and Middle East and reaches well into central Asia in the east. It extends as far north as the Altai Mountains (45°N) in the north and, uniquely, arises at the equator. There is a possibility that natural stands exist in southern Spain. It is most common in the Near East where valuable forest crops regenerate on the banks of rivers. Secondly, it has extremely curious foliage. Leaves on young, vigorous shoots on nursery plants tend to be narrowly lanceolate or linear. This juvenile, willow-type leaf may be found as well on established trees. Ordinarily, though, mature leaves are most often nearly circular or ovate or reniform (kidney shaped). Leaf margins vary from coarsely dentate to partly entire.

This wide distribution and highly variable foliage have led to many local forms being described by taxonomists and, in most cases, given species status. However, it is probable that all these putative taxa can be assigned to *P. euphratica*. The confusion has led to *P. euphratica* having at least 10 synonyms. More remarkably, its inclusion in the genus *Populus* has actually been questioned. The most radical arrangement has been put forward by Chardenon and Semizoglu (1962) who proposed the creation of a third genus – *Euphratodendron* – in the family *Salicaceae* and that *P. euphratica* should be re-named *E. olivieri*. The tree was often mentioned in the Bible as the willow of the children of Israel (Krussmann, 1986). It favours high temperatures and tolerates extremely saline soils.

In 1954 cuttings of *P. euphratica* were imported by the Forestry Commission from Baluchistan in West Pakistan but, on arrival, they were found to be too dry to insert. The following year, more material was obtained from Baluchistan and, in order to ensure successful propagation, scions were carefully selected and grafted on to understocks of *P. tremula*. Three grafts took and the resulting shoots were rather more than 1 m long by the end of the summer. In spite of the plants being provided with protection against frost however, none survived the first winter. There is no record that other attempts to reproduce this tree in Britain have been any more fortunate.

## Section Leuce, subsection Trepidae – *aspens*

### P. grandidentata Michx., large-toothed aspen

A native of north-east North America, where it is rarely taller than 20 m or has a diameter exceeding 60 cm. Little planted in Britain, there is no evidence to suggest that it could be successfully cultivated here. The few trees planted have not grown well, though this could reflect the use of poor provenances. It was used in important early breeding programmes in the section *Leuce* (see *P. grandidentata* × *alba*).

### *P. tremula* L., European aspen

This has an immense distribution, even greater than that of *Pinus sylvestris*, Scots pine. It is native to Britain and covers almost the whole of Europe and Western Asia, extending in the north and east through Siberia to the Bering Sea and in the south into North Africa. In Britain it is frequent only in north and west Scotland. Further south its occurrence is scattered and occasional, while in the south and east of England it is increasingly rare. In places it may go unnoticed. It is most commonly found in oak and birch woodland, and occasionally in pine woodland.

Aspen tolerates a wide range of soil types including heavy clay soils and freely drained sands and gravels. But even on fertile loams in the south of England it is one of the slowest growing and shortest lived poplars. On sheltered sites it may exceptionally reach a height of 20 m but, ordinarily, it is unlikely to be taller than 15 m. It is able to regenerate naturally from sucker shoots initiated by surface roots. In the most favourable situations dense, more or less pure aspen thickets can develop. On open ground, suckers may arise some 40 m away from the parent tree. Tree felling and soil disturbance stimulate sucker growth. Aspen is best known for its distinctive foliage. Because the leaves have a strongly flattened, relatively long petiole, they are constantly quivering and rustling even in the slightest breeze. They also colour well, being bronze in the early part of the growing season and then turning a clear yellow-gold at the end of the summer.

Most if not all provenances of aspen in this country are susceptible to bacterial canker. Near to persistent sources of infection they are among the first trees in mixed poplar to become infected and be severely damaged. It is possible that the local decline of aspen in some parts of the country may be due almost wholly to attack by bacterial canker.

Regrettably, aspen is rarely planted nowadays so that losses, whether due to bacterial canker or caused by other biotic factors, are not being replaced. Planting stocks are difficult to locate in any case, largely because of propagation problems. Fertile seeds are hardly ever found, hardwood cuttings cannot be rooted in open nursery beds, while plant production from suckers is uncertain and can only be practised on a small scale. Rooting softwood cuttings under mist is the best means of raising stocks.

Regardless of its native tree status and decreasing numbers, it deserves more attention than it receives. It is an attractive tree and tolerates a wide range of site conditions including salt laden winds at the seaside, and on these grounds merits inclusion in amenity plantings. A light demanding tree, though better able to put up with shade and competition than most other poplars, it is best planted at the edge of or in large gaps in woodland. Aspen is listed along with a few other native broadleaved trees for planting in clearings for browse for deer (Prior, 1983). It is much preferred by red and roe deer. Though it makes fine timber well suited for rotary peeling and the manufacture of excellent veneers and matches, there are no prospects of it being planted to produce merchantable wood because of low timber yields.

At least three geographical varieties of aspen have been described. Two of them, *P. tremula* var. *globosa* Dode and *P. tremula* var. *davidiana* (Dode) Seheid, an atypical form in China, are sometimes treated as species. The third variety, *P. tremula* var. *villosa* (Lang) Wesmael, a lowland type in western and southern Europe, is similar in certain respects to forms occurring in Britain. At least three horticultural cultivars are propagated – 'Pendula', 'Erecta' and 'Purpurea' – but all are rare. Only the first, weeping aspen, is found in Britain. A pendulous form, *P. tremula* var. *pendula* Jaeg., has been described in the wild.

### *P. tremuloides* Michx., quaking aspen

Sometimes called American aspen in European literature, this species is probably the most widely distributed tree of North America covering most of the continent with the exception of the central and south-eastern plains and tropical regions. It can reach a height of 30 m. It displays considerable variability and several varieties have been described in the wild. A pendulous cultivar 'Parasol de St. Julien' was selected in France more than 100 years ago. Quaking aspen has only been planted in Britain in gardens and arboreta. None of the specimens appears to have grown well or reached a large size, perhaps because of poor choice of provenance, and horticultural interest is now limited to the weeping cultivar. Interest in *P. tremuloides* in Britain and elsewhere in Europe arises because of its use as breeding stock for crossing with other species, especially *P. tremula* (see Hybrid aspens).

Two other species in this sub-section, *P. adenopoda* Maxim., Chinese aspen, and *P. sieboldii* Miq., Japanese aspen, are hardly ever planted in Britain, even in arboreta.

## Hybrid aspens

The best known aspen hybrid, at least in Europe, is between the two major species *P. tremula* and *P. tremuloides*. The first substantial, artificial breeding was carried out in the 1940s, in Denmark and Sweden, and by the end of the 1950s the hybrid was being produced in almost every European country including Britain. Like most other inter-specific first generation tree hybrids, *P. tremula* × *tremuloides* grows vigorously in both nursery and field. In this country the cross has proved to have much faster early rates of height and radial growth, and considerably straighter stems, than the native aspen. Unfortunately, though progeny from several different sources have been planted here including stocks bred at Alice Holt, most trees have been severely damaged by bacterial canker within a few years of planting and few have survived to reach a large size. Commercial planting of *P. tremula* × *tremuloides* is presently confined to Scandinavia where there is no bacterial canker. If disease resistant crosses could be bred there is little doubt that they would be widely planted in other parts of Europe. Breeding has so far been dependent upon imports of *P. tremuloides* pollen and the ready availability of suitable seed bearing *P. tremula*.

Other inter-specific aspen hybrids have been bred but none has attained the international prominence or the commercial potential of *P. tremula* × *tremuloides*. Perhaps the most interesting cross, at least from the horticultural point of view, is between *P. grandidentata* and *P. tremula*; it is the only aspen, species or hybrid, to be propagated readily from hardwood cuttings.

## *Section* Leuce, *subsection* Albidae – *white poplars*

### *P. alba* L., white poplar, abele

A native of central and southern Europe extending eastwards into western Siberia and central Asia and south into North Africa, it was an early introduction into Britain and has become naturalised locally. An important source of industrial wood in parts of the Near East, where it is grown commercially in close spaced plantations, it is planted only for amenity in this country and elsewhere in Europe. In Britain it is one of the slower growing poplars and its height scarcely ever exceeds 20 m. Several clones appear to be in horticultural use differing primarily in stem and crown shape. The most common has a twisting and leaning stem and the upper crown nearly always leans to one side. White poplar is often confused with *P.* × *canescens*, grey poplar, but this is faster growing and potentially much longer lived and larger tree, with a more attractively shaped crown. The leaves of white poplar are usually deeper lobed and more palmate than those of grey poplar, and the dense, wool-like hairs on young shoots and the under-surface of leaves of white poplar are a vivid, pure white compared with the pale greyish pubescence on the under-side of grey poplar leaves. White poplar tolerates salt-laden winds and exposure at the seaside and is therefore much planted at the coast in windbreaks, gardens and hedgerows. It grows well in coastal sands and gravels, and because sucker growth can develop at an early age bare areas may be quickly colonised. It is also planted frequently at the road side, on reclaimed industrial sites and on open ground in housing and industrial estates, usually in intimate mixture with other broadleaved trees. In gardens and parks its suckering habit lessens its value. Leaves colour well, though briefly, in the autumn, generally becoming bright gold-yellow or, much less often, crimson. Several clones including both male and female forms are cultivated in Britain; all are readily propagated from hardwood cuttings in open nursery beds. There is no evidence to suggest that any are seriously susceptible to bacterial canker. Several varieties of *P. alba* have been described and some attractive ornamental cultivars are used in horticulture. Three deserve to be planted in Britain's parks and gardens, namely 'Pyramidalis', 'Raket' and 'Richardii'.

*P. alba* 'Pyramidalis' or Bolle's poplar is a male cultivar sometimes named *P. alba* 'Bolleana' or *P. bolleana* Lauche in nursery stock lists, it has a semi-fastigiate, pyramidal crown quite unlike that of white poplar, which it can exceed in size, and it nearly always has a straight and upright stem. In other respects it is like *P. alba*. It is moderately susceptible to bacterial canker and trees close to sources of infection can be so seriously damaged that dieback and even death can result. But in disease-free localities trees may reach sizeable dimensions (23 m × 85 cm diameter at 1.3 m). The cultivar was introduced to western Europe from Uzbekskaya, 150 miles north of the USSR-Afghanistan border as long ago as the late 1870s.

*P. alba* 'Raket' is a hybrid (*P. alba* × *P. alba* 'Pyramidalis') bred artificially in the Netherlands, where it was released in 1972. It has a narrow, fastigiate crown and a low susceptibility to bacterial canker and foliar diseases. It is recommended in the Netherlands for planting in gardens, parks and windbreaks. A 30-year-old specimen tree in the National Populetum at Alice Holt Lodge, raised from the same cross carried out in Hungary, has a similar crown and stem form to that of most white poplars in Britain and shows few characteristics of *P. alba* 'Pyramidalis'.

*P. alba* 'Richardii' was selected in the Netherlands in 1918 for its decorative foliage. The leaf undersurface is typically white while the upper surface is a bright gold-yellow or light lime green. It is increasingly rare in Britain.

### *P. tomentosa* Carr., Chinese white poplar

A native of northern China and much planted everywhere in that country especially in and around Peking. It has been little planted in Europe. In Britain the number of true-to-name specimens is unknown but it is unlikely to be greater than one or two. A tree planted in 1959 in the National Populetum at Alice Holt Lodge may be authentic. It was raised from a cutting supplied as *P. tomentosa* in 1955 by the Ontario Forest Research Centre, Maple, Ontario, Canada, from stock derived from a tree at Arnold Arboretum, Jamaica Plain, Massachusetts, USA. The Forestry Commission tree has a dense, heavily leaved, rounded crown with many pendulous branches and is one of the handsomest poplars in the collection. Its foliage and suckering habit are similar to that of grey poplar and it is as difficult to propagate from hardwood cuttings as this tree. But softwood cuttings taken in midsummer root under mist within two weeks of insertion. Chinese white poplar is classified with *P. alba* but it is closer botanically to grey poplar and, like this tree, may be a hybrid. If so, its probable parents are *P. alba* and *P. tremula* var. *davidiana*.

## Hybrids between subsections Trepidae *and* Albidae

### *P. grandidentata* × *P. alba*

A hybrid produced artificially since the 1930s, mainly in Canada and Hungary. Three clones bred in Canada were planted in Forestry Commission trials in the 1950s. On most sites they grew rapidly to begin with and, for several years after planting, appeared to have commercial potential. The fastest growing trees were 15 m tall at 11 years of age. However, in every trial, even in sheltered localities, the trees were heavily branched and had a poorly shaped stem and crown. Thus in spite of their early vigorous growth none of the clones was thought to be sufficiently attractive to justify further planting. Breeding has continued in Canada and Hungary, and hybrids have been produced suitable for growing in forest conditions.

### *P.* × *canescens* Smith, grey poplar

The status of this tree in Britain is uncertain. While it is thought by Clapham *et al.* (1962) to be probably native in moist lowland woodland in southern England, other authorities are less committed. It is locally common throughout the country, especially between the north Midlands and Dorset, and in many places is naturalised. It has a wide distribution in Europe and Western Asia and extends into the south of the USSR. It is a hybrid between *P. tremula* and *P. alba*, and all trees display characteristics inherited from both parents. Where the two species, represented by trees of opposite sex, occur together hybrids between them may arise spontaneously. Grey and white poplar have similar foliage and the two trees are frequently confused (see p. 7). But grey poplar is by far the more handsome tree, achieving much larger dimensions than white poplar (up to 35 m), and should be preferred. It is also the faster growing and longer lived of the two.

Grey poplar tolerates a wide range of soils including moist, heavy clay soils and both acid and alkaline soils so long as they remain damp during the summer. It is not successful on dry, heathland soils. It grows well near the sea and can be cultivated as far north and east as Sutherland. It is best as a specimen tree so long as space is available, so that its silvery foliage, tall stature, fine bole and high domed crown can be properly appreciated. But it succeeds in woodland and can quickly colonise ground by suckering to form small, pure stands. It has been recommended with several other trees for replacing elm in the countryside (Mitchell, 1973).

Grey poplar is impossible to propagate from hardwood cuttings in open nursery beds. However, softwood and semi-ripe cuttings taken at any time during the growing season root quickly under mist. Softwood cuttings inserted in early summer are rooted after only two weeks and, following a brief period of weaning, may then grow fast enough to reach a height of 1 m by the autumn, whether potted-up or bedded-out. Stock for field planting can also be produced from carefully lifted suckers. In Germany, several clones of grey poplar have been given cultivar names and released to horticulture.

*P. tremula* and *P. alba* have been crossed artificially in several Western European countries. Hybrids are easily produced. Much of the breeding has been done in Germany where clones have been selected and propagated for field trials.

Many other hybrids between *Trepidae* and *Albidae* poplars have been produced but few have achieved prominence. The Asian species *P. adenopoda* and *P. tomentosa* have sometimes been used as parents. An improvement in the rooting ability of the aspens has been one of the aims of the work. Only one cross has attained commercial importance. *P. glandulosa* × *alba* hybrids now cover well over 10 000 ha in Korea.

## Section Aigeiros – *black poplars*

World wide more than 90 per cent of all cultivated poplars are in the section *Aigeiros*. In North America black poplars are primarily pioneer trees readily colonising valley bottoms, especially alluvial soils after flooding. Large, highly productive and commercially important stands yielding valuable merchantable timber can develop. In contrast, planting of black poplars in North America started only two to three decades ago and is still relatively unimportant. By comparison, natural stands of black poplar in Europe and Asia are usually of limited economic value but they are extensively planted in many parts of both continents. In central and southern Europe planted crops of black poplar – almost exclusively *P.* × *euramericana* hybrids – cover more than one million hectares.

### *P. deltoides* Marsh, eastern cottonwood

A native of eastern North America where it regenerates freely from seed to form productive woodland. Heights in excess of 30 m can be achieved. There are three important sub-species. A large number of clones have been selected for nursery and field trials and some have been described botanically. *P. deltoides* was introduced into Europe towards the end of the 17th century though early planting was limited to a few vegetatively raised clones. The best known of these are named 'Carolin' and 'Cordata'. More recently large quantities of seed of carefully selected provenances have been imported from North America by European research workers and during the past few years, after vigorous testing of progeny in field and disease trials, clones have been selected, described and released to horticulture under registered cultivar names. The best known are 'Alcinde', a French selection, 'Lincoln', 'Marquette' and 'Peoria', released in

Germany, and 'Harvard', 'Lux' and 'Onda', selections made in Italy. *P. deltoides* has been little planted in Britain and is now only seen in a few arboreta. There is not a convincing argument for cultivating it in this country; many other species and cultivars, including *P. nigra* for amenity and *P. trichocarpa* and its hybrids for wood production, are far more suited to our somewhat difficult climate and soils.

### *P. fremontii* S. Wats., *P. sargentii* Dode and *P. wislizenii* Sarg.

These are black poplars locally important in the USA but, seemingly, with little value in Europe. They are seen only in clonal collections and a few arboreta.

### *P. nigra* L., black poplar

An extremely variable tree with a wide distribution in Europe, western and central Asia and North Africa. A variety, *P. nigra* var. *betulifolia*, is considered to be a native of Britain and, in turn, is also very variable. A cultivar of black poplar, *P. nigra* 'Italica', Lombardy poplar, is one of the best known and most widely planted trees in the world. To most people it epitomises all poplars. Many forms of black poplar develop large burrs on the surface of the stem. It is quite probable that several black poplar cultivars are grown in different parts of the world under a variety of names, some often quite local. The species is considered to be highly resistant to bacterial canker. A brief review of the main varieties and cultivars follows.

### *P. nigra* var. *betulifolia* (Pursh) Torr., downy black poplar

A western form of *P. nigra* which occurs naturally in north-west France and is considered a native of Britain although there is little current evidence to suggest that many of the specimens still surviving here can have arisen spontaneously from seed. It has pubescent petioles, midribs and young branchlets in contrast to the glabrous type. But over much of the natural range of black poplar, pubescent and glabrous forms can be found growing side by side. It occasionally reaches a height of 30 m. In Britain black poplar displays considerable variation in macrocharacteristics. For instance, obvious differences in crown and stem shape, in relative number of stem burrs and in date of leafing can be observed from tree to tree. Both male and female specimens are found; male trees predominate. A male tree called Manchester poplar has been widely planted in towns in the Midlands and the north of England on account of its high resistance to atmospheric smoke pollution. It is the most heavily burred black poplar and often has a rounder and more compact crown than other forms. It appears to be a single clone or a mixture of a small number of virtually identical clones. Male specimens of *P. nigra* also abound in the Vale of Aylesbury, Buckinghamshire. Here, however, the trees show appreciable variation and reflect several different, unknown origins. An appreciable number of specimens are quite dissimilar from Manchester poplar, lacking or having few stem burrs and being open-crowned. They are also later into leaf than Manchester poplar.

In most parts of Britain black poplar is becoming rare and perhaps endangered. During the past 30 years or so large numbers of old trees have been felled and, regrettably, few have been replaced. An even greater number of specimens are fast approaching maturity – some are already in poor health – and there is little sign that after felling these will be replaced either. Fortunately botanists and conservationists are concerned about the declining black poplar population, and some steps have been taken with the help of the media to ensure that selected old trees are properly looked after and that planting stocks of black poplar are available in nurseries to replace lost specimens. A survey of black poplar in Britain was launched in 1973 by the Botanical Society of the British Isles; several accounts of its aims and progress have been published by Milne-Redhead (1984, 1985), who is responsible for the collection and collation of field evidence.

### *P. nigra* var. *caudina* Tenore and *P. nigra* var. *neapolitana* Tenore

These are varieties described in southern Italy and, unless exceptional in some respect, are unlikely to be cultivated in Britain.

### *P. nigra* var. *thevestina* Dode

A variable fastigiate tree covering much of the Near East and extending into North Africa. Several clones have been described and registered by the International Poplar Commission. The best known is 'Hamoui', a cultivar widely planted in Iran, Iraq, Jordan, Lebanon and Syria, sometimes under other names, as well as in south-east Europe. It is nearly always cultivated at close spacing, whether in plantations or in single or double lines. It is a remarkable tree in two respects; it is well adapted to a hot, dry climate and it has one of the most beautiful barks – usually a shiny silvery-white or grey until old age – of any poplar. It is a female tree.

### *P. nigra* 'Charkowiensis'

Believed to have arisen as a naturally regenerated seedling in a nursery near Kharkov, USSR, towards the end of the 19th century. It was cultivated subsequently in western Germany though only for a relatively short time and hardly at all elsewhere. It is a female tree. Specimens in the National Populetum at Alice Holt Lodge, one raised from material imported in the mid 1930s from Hamburg, Germany, and planted in 1954, the other propagated from cuttings obtained in 1957 from Kursk Province, USSR, via the Academy of Sciences, Moscow and planted in 1960, have a narrow though not fastigiate crown and an unburred much lighter coloured bark than is usual in *P. nigra*. The cultivar is now only of botanical interest.

### *P. nigra* 'Chile'

A cultivar grown in Argentina and Chile, often in association with *P. nigra* 'Italica' and other fastigiate poplars. Remarkable in South America for its almost evergreen habit, it was imported by the Forestry Commission from Argentina in 1953 under the name *P. nigra sempervirens* 'Chileno'. Although stools and trees were successfully propagated at Alice Holt Lodge for several years, and proved to hold their leaves well into the winter, the cultivar was extremely prone to frost injury and all stocks eventually died in the 1960s. In contrast, a clone obtained in 1950 from Argentina, under the name *P. nigra* 'Criollo', has been maintained successfully at Alice Holt Lodge since its introduction. Trees planted in the National Populetum in 1955 remain healthy and apparently free of winter cold damage. It is a female clone and chiefly of interest to growers in Britain as it is identical in most if not all respects to *P. nigra* 'Gigantea', the so-called female or giant Lombardy poplar which has been grown here for 100 years.

### *P. nigra* 'Gigantea', female or giant Lombardy poplar

A female, fastigiate cultivar closely resembling *P. nigra* 'Italica' but with a broader and looser crown. Its origin is unknown though it was almost certainly received at Kew as long ago as 1880. Sometimes found as a specimen tree, it is usually seen in mixture with Lombardy poplar in lines or groups of trees supposedly composed only of this cultivar. Occasionally a third fastigiate clone, *P. nigra* 'Plantierensis', may be present in the mixture. Where 'Gigantea' and 'Italica' are planted together, the former is probably there by accident as a rogue. 'Gigantea' is a more than adequate substitute for 'Italica'. It is much less prone to fungal diseases and to dieback and death of branches than Lombardy poplar and is at least as vigorous as this cultivar.

### *P. nigra* 'Italica', Lombardy poplar.

Planted all over the world for ornament, screening and shelter and sometimes for timber production. It is the best known and most commonly planted fastigiate tree. It is a male tree bearing numerous red catkins in late March, it tolerates a wide range of soils and climates and grows rapidly for several years after planting, though not as fast as most other common poplars. In favourable localities it can reach a height of 35 m – the tallest specimen in Britain is 38.5 m – but most trees are much shorter than this. It is best planted in groups or as a single specimen to contrast with broad-crowned trees, and the temptation to establish it in long lines at the boundary of fields or gardens, or to screen factories, quarries and spoil heaps should be avoided. Fortunately it is much less planted nowadays than in the past. Even so it is still planted needlessly when other trees would seem more appropriate.

Lombardy poplar is highly susceptible to *Marssonina populi-nigrae*, a leaf fungus which causes foliage to wither prematurely. Infection starts at the base of the crown and progresses upwards; by the end of the summer badly attacked trees may be almost wholly defoliated, at best retaining leaves only in the top metre or so of the crown. *Marssonina* defoliation also renders branches susceptible to attack by weak bark pathogens like *Cytospora chrysosperma* (Strouts, 1980). Invasion by *Cytospora* leads to canker formation and death of patches of bark. Lower branches are attacked first. Death of branches may follow and in certain circumstances trees may be killed. The dire consequences of *Marssonina* attack on Lombardy poplar can be seen in town and country year after year. Two other strongly fastigiate cultivars, *P. nigra* 'Gigantea' and *P. nigra* 'Plantierensis', can be safely used as substitutes for this disease prone poplar, however. Both retain their leaves well into the autumn.

The origin of Lombardy poplar is unknown. But in the mid-18th century it was quickly spread from the Po Valley to other parts of Europe and it is fashionable to accept this area in northern Italy as its source. Not all poplar experts agree with this view.

## *P. nigra* 'Plantierensis'

A fastigiate male cultivar closely resembling Lombardy poplar, to which it is preferred. It arose in a nursery at Metz, north-east France, and was first propagated and planted there towards the end of the 19th century. It is probably a cross between *P. nigra* var. *betulifolia* and *P. nigra* 'Italica'. 'Plantierensis' and 'Italica' are easily separated botanically; the former has pubescent branchlets and petioles, a characteristic best checked before the end of the summer, while in contrast 'Italica' is wholly glabrous. Where the two cultivars occur together in the same planting, a common enough event, the relative number of leaves in the crown is the easiest means of identification. Unlike Lombardy poplar, 'Plantierensis' always has a dense, well-leaved crown and invariably retains its foliage into the autumn. For these reasons it is much the better prospect of the two when a fastigiate poplar is required. Stocks of 'Plantierensis' should, of course, continue to be raised in nurseries under that name and no attempt should be made to describe them as Lombardy poplar.

## *P. nigra* 'Vereecken'

A male clone cultivated in the Netherlands, where it was released to the nursery trade in 1959. It is thought to have originated in Belgium. It was imported by the Forestry Commission in 1950, and in 1954 it was planted in the National Populetum where it has proved to be the fastest growing black poplar clone. A specimen which was 19.5 m tall and had a breast height diameter of 36 cm at age 20 had reached a height of 24 m and a diameter of 60 cm after 32 years (1985). It has a columnar, comparatively narrow crown with ascending branches and its straight stem is unburred. It has so far remained free of disease. The cultivar is highly thought of in the Netherlands where it has an excellent record of resistance to bacterial canker. It deserves to be planted in Britain in parks and large gardens.

Six other *P. nigra* cultivars have been released recently in the Netherlands and all are showing considerable promise. These are more likely to be cultivated in Britain in the future than any of the lesser known black poplar clones selected primarily in France, Germany and Italy towards the end of the last century or shortly afterwards.

## *P.* × *euramericana* (Dode) Guinier, black poplar hybrids

A very large group of natural and artificial hybrids derived from *P. deltoides* and *P. nigra* and often referred to as hybrid black poplars. The occurrence of natural hybrids between the two species was recorded in Europe as long ago as the early 1700s soon after the introduction of *P. deltoides* from North America. The most important hybrid progeny arose in the major river valleys of central and southern Europe, where *P. nigra* was already being extensively cultivated and where conditions favoured successful seed germination, seedling survival and fast rates of growth. Sometimes, complex populations evolved due to backcrossing and hybridisation between inter-specific first generation crosses.

On the whole the hybrids were seen to be faster growing than either the native *P. nigra* or the introduced *P. deltoides*, and outstanding specimens were soon selected for propagation and local cultivation. By the end of the 1700s a few of the new clones were being cultivated nationally and, in one or two instances, in other countries. Some have remained in commerce and are still being planted today.

Towards the end of the 18th century the preparation of botanical descriptions and the provision of formal nomenclature for poplars had begun to preoccupy taxonomists.

Interest in natural black poplar hybrids probably reached a peak in the 1930s and 40s when research programmes were started in almost every European country to study and compare clonal behaviour. During the past three decades interest in the selection and testing of natural *P. × euramericana* hybrids has greatly declined in Europe, although some important investigations have continued. Instead, resources have been switched to artificial breeding programmes between *P. deltoides* and *P. nigra*. Modern breeding techniques and improved scientific methods for examining clones for resistance to disease, as well as the use of carefully selected and well-tested *P. nigra* parents and the increasing availability of cuttings and seed of *P. deltoides* have ensured rapid progress and the continuing production of potentially high quality hybrid progeny. Since the programmes started extremely large numbers of seedlings have been produced, in total perhaps in the region of two million.

Black poplar hybrids were formerly classified under the name *P. canadensis* Moench. This nomenclature, proposed in 1785, probably to describe *P.* 'Serotina' caused much confusion, however, sometimes being applied to North American species and in some instances to any hybrid. Moreover the name implies, wholly incorrectly, a specific geographical origin, while the hybrid origin of the poplars Moench described was not defined in the botanical accounts. To resolve these taxonomic anomalies and to prevent confusion in future, the International Poplar Commission decided in 1950 to reject *P. canadensis* and to substitute in its place the collective name *P. × euramericana* (Dode) Guinier. This nomenclature, with a Latin description, was proposed to the International Botanical Congress in that year. It was further decided by the Commission to retain the scientific names of the old hybrids but to give them cultivar status. Since then the formal nomenclature for all black poplar hybrids has been *P. × euramericana* 'Serotina' (or cv. Serotina), for example, with *P.* 'Serotina' as the shortened version.

Black poplar hybrids attain their highest growth rates in southern Europe. At close spacing rotations of 8 to 10 years for pulpwood are commonplace and, at wide spacing, sawlogs and veneer logs are expected at 13 to 15 years of age. In Britain, and elsewhere in northern Europe where the growing seasons are cooler and shorter and light intensities are lower, growth rates are correspondingly reduced. In this country they are consistently vigorous only in the southern half of England, although in northern England, Wales and Scotland they can be successfully cultivated on better soils in relatively mild, sheltered localities – but on longer rotations.

In Britain early introductions of *P. × euramericana* cultivars included many since found to be unsuitable because of slow growth rates or on account of susceptibility to bacterial canker. Little is gained by discussing these inferior poplars and comment is therefore limited in the review that follows to cultivars that have proved their worth in the long term or which may merit cultivation in future.

## *P.* 'Eugenei'

Arose spontaneously in a nursery near Metz in north-east France in 1832 and was imported into Britain in 1888. It is cultivated only in this country. Lombardy poplar was probably the male parent and *P.* 'Marilandica' perhaps the female. A male tree, it has a comparatively narrow, symmetrical, ascending crown and a remarkably unforked, straight stem. It grows particularly well in the south and east of England and can succeed on sites ordinarily considered too dry for poplar. It is one of the best cultivars for open, exposed but not upland sites, since its crown and stem shape are little affected by wind. As a consequence it has been used in single-row windbreaks. It has been grown successfully in the Fens for many years to produce veneer quality wood for the manufacture of chip baskets. In favourable localities it can easily reach a height of 30 m in 25 to 30 years. A specimen at Colesbourne, Gloucestershire, was in 1985 the fourth tallest broadleaved tree in Britain (Mitchell and Hallett, 1985). At 68 years of age it was 38 m tall with a breast height diameter of 111 cm; 14 years later when 82 years old (1984) it had reached a height of 43 m and a diameter of 135 cm and showed no signs of stopping. It is usually faster growing than the more commonly planted 'Robusta' and 'Serotina'.

In 1965 it was put forward by the Forestry Commission for registration with the International Poplar Commission, when a formal description was lodged with the Commission's secretariat at the Food and Agricultural Organization of the United Nations in Rome. It has a good record of resistance to bacterial canker and is never seriously infected in the field. It is, however, susceptible to *Marssonina brunnea*, a leaf fungus causing premature defoliation and loss of increment and, therefore, can no longer be recommended for planting.

## P. 'Florence Biondi'

A Stout and Schreiner hybrid bred in the USA in 1925 and imported into the Netherlands 20 years later as OP 226 or NE 226 where it was released to growers in 1972 as 'Florence Biondi'. OP 226 was tested at Alice Holt Lodge in the 1950s and found to be a vigorous cultivar with good stem and crown form, resistant to *Marssonina populi-nigrae* and to bacterial canker. However, a second batch of cuttings imported from the Netherlands was found to be susceptible to canker and later it was discovered that two almost identical clones exist. Consequently OP 226 or 'Florence Biondi' has not been released in Britain.

## P. 'Gelrica'

A male cultivar selected in the Netherlands in 1865 and widely planted in that country until recently. It has been little planted elsewhere, but following its introduction into Britain in 1937, the year of its official release in the Netherlands, it was included in Forestry Commission trials and on a small scale in private sector plantings until the 1960s. It has the lightest coloured and perhaps most attractive bark of any black poplar hybrid but is otherwise unexceptional in appearance. It is similar in habit to *P.* 'Serotina' and on exposed sites may sometimes lean with the prevailing wind, though never as badly as 'Serotina'. In the past, serious fluting of stems has been observed, though fortunately in relatively few crops, and this defect has lessened prospects of the logs being rotary peeled for veneers. The cause of the disorder, which appears to be peculiar to 'Gelrica', is unknown.

'Gelrica' is moderately resistant to bacterial canker and in practice infection is rare. If damage occurs it tends to be slight, sporadic and confined to twigs and branches. Its susceptibility to *Marssonina brunnea* is a greater cause for concern, though in Britain there is no record of this leaf fungus causing lasting damage. But in the Netherlands serious outbreaks have occurred and repeated attacks have led to the death of large numbers of trees. The worst damage occurred in the 1960s and it is very probable that the marked decline in planting rate of 'Gelrica' in the Netherlands in recent years is attributable almost wholly to its susceptibility to *Marssonina*.

'Gelrica' is chiefly noted for its rapid rate of radial growth. In fact, on poplar sites throughout Britain its diameter increment almost always exceeds that of any other black poplar hybrid. Not surprisingly it is most vigorous in fertile soil in the southern half of England, where stems pruned to 6 m can reach rotary veneer specifications in 15 to 20 years. Trees with a breast height diameter of 60 cm at 20 years are not uncommon and a diameter of more than 80 cm can be attained in 35 years. Because high yields are assured on most sites in Britain 'Gelrica' is approved by the Forestry Commission for planting for wood production.

## P. 'Heidemij'

A male cultivar selected in the Netherlands and first propagated there in 1891. It was officially released in that country in 1937. Until 1957 it was cultivated under the name *P. deltoides missouriensis* and was thought by some authorities to be a variety of eastern cottonwood and to have an American origin. However, it is clearly a euramerican hybrid.

It was introduced into Britain in 1950 and has been recommended ever since for planting for wood production, though it is still rarely chosen. It shows good resistance to bacterial canker and is not prone to attack by *Marssonina brunnea*. But it is very susceptible to *Melampsora* leaf rusts and serious infection can lead to premature leaf-fall and loss of increment.

'Heidemij' has the same habit, growth rate and botanical features, with one notable exception, as *P.* 'Robusta' and is indistinguishable in appearance from that widely planted cultivar. The only reliable means of separating the two poplars is by determining with a good lens the occurrence of minute erect hairs on young branchlets and petioles. While 'Robusta' always has downy young shoots and petioles, those of 'Heidemij' are always glabrous.

A male poplar imported into Britain from France in 1916 under the name *P.* 'Laevigiata' and included later in Forestry Commission trials was found in the 1960s, after careful examination of botanical characters, to be identical with *P.* 'Heidemij'. As a consequence a recommendation was made to give the name *P.* 'Heidemij' to all nursery stocks and trees then labelled *P.* 'Laevigiata'.

## *P.* 'I-78'

This is a female cultivar imported into Britain in 1939. It arose spontaneously in Italy, probably in the Po Valley, and was selected and first propagated in about 1929. Like a large number of other hybrids derived in Italy from *P. deltoides* × *P. nigra* which have since been cultivated throughout the world, it was tested at the Poplar Research Institute at Casale Monferrato.

'I-78' grows poorly in northern Britain and is therefore recommended for use only in the southern half of England, the only poplar to be so limited. However, on selected sites it grows as well as or rather better than *P.* 'Robusta' and trees large enough for rotary peeling can be obtained in 20 to 25 years. On sheltered ground its height can easily exceed 30 m in 35 years. Its ultimate dimensions can only be guessed at since the earliest planted trees in Britain are still less than 40 years old.

Though not resistant to bacterial canker, 'I-78' has been little damaged in trials even where the disease was artificially introduced. Where natural infection has occurred, the attack has usually been limited to small twigs. It is moderately susceptible to *Marssonina brunnea* and serious infection can cause premature leaf fall and loss of increment. It usually has a straight and vertical stem and a crown becoming broader with age but remaining symmetrical, 'I-78' is also cultivated in Germany.

## *P.* 'I-214'

This is a female cultivar, and is much the best known Italian black poplar hybrid. It was selected and placed under trial at about the same time as 'I-78'. But unlike 'I-78', which is grown in only two countries, both in north-west Europe, 'I-214', is cultivated all over the world. It is a highly adaptable and extremely vigorous tree, often achieving growth rates and yields of merchantable wood far in excess of those of other poplars. Rotations of 15 years or less for the production of sawlogs and veneer logs are commonplace. It has proved most successful in countries enjoying long, warm summers, encouraging the establishment of extensive plantations. Yet it has several drawbacks and requires attentive management. A tendency to develop large branches, sometimes leading to serious forking, and moderate susceptibility to *Marssonina brunnea* are the best known defects. It is markedly phototropic, so that trees growing in conditions of partial shade will grow strongly towards the light. In Britain, it has proved successful only on sheltered, fertile sites in southern England. By and large its behaviour has been too unpredictable to warrant extensive planting. A proneness to develop bends in the lower stem has also greatly lessened its potential usefulness in this country.

## 'I-45/51', 'I-154', 'I-262' *and* 'I-455'

These are other Italian hybrids cultivated world

wide. Although clones released in Italy in recent years have all been given cultivar names, none of the earliest selections was named, other than 'I-154'. In its early days it was called 'AM' (Arnolos Mussolini).

## P. 'Marilandica'

This is a female cultivar believed to have arisen spontaneously at the end of the 18th century. Its parents are considered to be *P. nigra* (female) and *P.* 'Serotina'. Its nomenclature suggests an origin in the United States of America but it is undoubtedly a European hybrid. For a long time it was called the Canadian poplar, further confusing its source. Though little planted nowadays, it was formerly cultivated in several countries in western and central Europe to produce a high quality timber. It was much favoured in Germany and the Netherlands. It was introduced to Britain in 1843 and first planted three or four years later at the Royal Botanic Gardens, Kew. The oldest specimen at Kew, planted around 1846, remains in good health and continues to increase in size. In 1974 it had a height of 35 m and a breast height diameter of 164 cm; in 1984 its height had increased to 37 m and its diameter to 173 cm. Though only slightly susceptible to bacterial canker it has never been a popular choice in this country either for timber production or for amenity. During the past few decades it has hardly been planted at all and is now becoming quite rare as old trees mature and are felled.

## P. 'Regenerata'

Several female clones have been cultivated under the name 'Regenerata'. Most if not all appear to have arisen spontaneously, in France, the earliest probably from 'Marilandica' (female) and 'Serotina'. The first selection was made in 1814, at Arcueil in the south of Paris, only 5 miles from the city centre. It was subsequently cultivated extensively in France and then all over Europe. 'Regenerata' was introduced to Britain in about 1870 and became one of the most widely planted poplars until succeeded in the 1940s and 50s by more reliable cultivars. Though grown in plantations to produce merchantable wood, it is best known in this country for its use in screens and shelterbelts. Remnants of 'Regenerata' screens are still common in the south and east of England and, perhaps paradoxically, are more likely to be seen in urban localities than in the countryside. 'Regenerata' was at one time much favoured for planting along railways and in goods-yards and sidings prompting Mitchell (1974) to name 'Regenerata' railway poplar. It is susceptible to bacterial canker and should no longer be planted.

## P. 'Robusta'

A male cultivar which arose spontaneously in a nursery near Metz in the north-east of France in about 1890. *P.* 'Eugenei' originated in the same nursery some 60 years earlier and was thought at first to be the male parent of 'Robusta'. However, this is no longer considered likely and the identity of the pollen-parent remains unknown. It is presumed to be a glabrous form of *P. nigra*, though, and Bean (1976) goes so far as to suggest that it was most probably *P. nigra* 'Plantierensis'. This fastigiate cultivar arose at the same nursery some years before 'Robusta'. The female parent was *P. deltoides angulata*.

'Robusta' is grown all over western Europe and elsewhere in the world wherever poplar culture is important. In several European countries it has accounted until recently for more than half the poplars planted. Sometimes it is as common in line plantings as in blocks. Introduced to Britain in about 1910 it was confined for a time to a few gardens. But by the end of the 1920s it had been successfully established in single rows, avenues, small plantations and groups in woodland, mainly at Ryston Hall, Norfolk, in the expectation of producing merchantable wood. One of the more interesting stands at Ryston Hall planted in 1928 was a mixture of 'Robusta' and Lawson cypress, *Chamaecyparis lawsoniana*. In the 1930s 'Robusta' was included in early Forestry Commission trials with other newly acquired cultivars. Also it was planted for the first time at the roadside for amenity and in single row windbreaks to protect valuable farm crops. Planting reached a peak in the 1950s and 60s when large numbers of poplar stands were established for the production of timber. During these two decades more than half the trees planted were 'Robusta' and in most seasons the area of 'Robusta' planted exceeded 100 ha.

In most respects 'Robusta' is the handsomest euramerican cultivar. In the first place it is the earliest hybrid into leaf and for a week or two in late April–early May the young foliage, as it unfolds and grows, is a striking bronze–red. Later the leaves become attractively glossy and deep green. By the end of June, the leaves are fully developed, large and abundant throughout the crown. As a consequence the crown for much of the summer is more dense than in others in the group and gives better screening, sound and dust-filtering qualities than its relatives. Further, the crown remains comparatively narrow and columnar, and the stem straight and upright. The production of enormous numbers of long (6–9 cm) and deep brick-red catkins in early April, sometimes identifying trees at a long distance, is another fine feature. But 'Robusta' not only makes a shapely and colourful tree on a wide range of sites but it is also one of the fastest growing euramerican cultivars.

For a time a few years ago there were fears that 'Robusta' might be too susceptible to bacterial canker to permit its successful cultivation in all parts of Britain. This anxiety arose partly because of inexplicable canker damage in one or two timber crops and natural infection in trial plots close to sources of the disease or where the disease had been artificially introduced. There was also concern that the development of serious stem cankers in disease trials following inoculation with natural bacterial slime might indicate moderate to high susceptibility. In fact, considering its popularity for amenity and widespread use for timber production, very few serious canker problems have come to light. Though infection of twigs is not uncommon, erumpent cankers on branches, leading to girdling and dieback, are rare. Stem cankers causing loss of increment and degrade of timber are virtually unknown. Moreover, 'Robusta' is resistant to *Marssonina brunnea* and in several western European countries, where trees of other poplars have died due to repeated attack by this leaf fungus, interest in the cultivar has revived because of its freedom from *Marssonina* infection. But it is susceptible to one of the poplar leaf rust fungi, *Melampsora larici-populina*. Though the disease is primarily a problem in nurseries, serious outbreaks in the field can cause premature leaf-fall of adult trees and consequent loss of increment and die-back of the previous year's shoots. The worst damage to 'Robusta' by this fungus appears to have occurred in Belgium.

A new physiological race of *M. larici-populina* was thought to be responsible (Steenackers, 1982). Fortunately, infection in Britain usually occurs towards the end of the season and, while trees and nursery plants especially can be made unsightly for a short while, detectable permanent injury is rare.

## *P.* 'Serotina'

Usually regarded as the first spontaneous euramerican hybrid, it arose early in the 18th century, probably in France, and has been cultivated continually ever since in western and central Europe. A male cultivar and the last poplar into leaf, it was, for a long time, mistakenly called the Swiss poplar. It is sometimes considered to be a mixture of several identical clones. Introduced to Britain about 200 years ago, it was much favoured as an amenity tree in both town and country and, though not among the fastest growing poplars, many early planted specimens have been able to reach an impressively large size. Bean (1976) goes so far as to suggest that 'Serotina' attains greater dimensions than any other introduced broadleaved tree. Certainly in the 1950s several unusually big specimens were to be found in the south and east of England. Some were remarkably handsome, much approved of locally and dwarfing all other trees in the proximity, and photographs of them regularly appeared in books, magazines and calendars. The stoutest had stem diameters exceeding 2 m at breast height (1.3 m) and a few were estimated to have a total volume including branchwood of over 36 $m^3$ (1000 $ft^3$). A fine specimen at Bowood, near Calne, Wiltshire, is now the second tallest broadleaved tree in Britain according to Mitchell and Hallett (1985). In recent years, however, other notable 'Serotina' have, regrettably, blown over, broken up or had to be felled.

'Serotina' is resistant to *Marssonina brunnea* and only slightly susceptible to *Melampsora* leaf rust fungi. As a result, leaf retention is good at the end of the season. It is not resistant to bacterial canker but in practice serious infection is rare and usually confined to twigs and small branches. Its most serious defect is a tendency to lean with the prevailing wind. The greater the exposure, the more the lean and consequent reduction in timber quality. When crops are planted to produce merchantable wood, therefore, sheltered sites should be preferred.

For a time *P.* 'Serotina' was known in Britain as Canadian poplar. Nowadays it is called black Italian poplar, though this is an equally inappropriate and misleading name.

## *P.* 'Serotina Aurea', golden poplar

Arose in 1871 as a branch sport, in a nursery at Kalmthout in the extreme north of Belgium, and released to the horticultural trade in that country 5 years later. Its leaves remain a bright golden yellow for much of the growing season, only becoming greenish yellow at the end. It grows well in towns, and in large gardens and parks it contrasts attractively with other large trees. Though slower growing than the type *P.* 'Serotina', it is like this cultivar in crown and stem habit, resistance to disease, date of leafing and leaf-shape. It can easily reach a height of 30 m and is the largest deciduous tree with golden foliage. It deserves to be more widely planted.

## *P.* 'Serotina de Selys'

Arose in Belgium, perhaps as a mutation. It has been known since 1818. It is like 'Serotina' in most respects, differing from this mainly in crown shape and rate of growth. Because it has a fastigiate habit it is usually compared with and mistaken for the better known Lombardy poplar. Yet the two are easily separated on branch and crown features. 'Serotina de Selys' tends to have much shorter branches, of more uniform length than those of the Lombardy, and its crown is nearly always narrower and inclined to be cylindrical in silhouette. The two can also be readily separated on time of leafing, 'Serotina de Selys' flushing as much as three weeks after Lombardy poplar. Its late leafing is perhaps its only defect. It is a rare tree in Britain and deserves to be planted more often. In parks and gardens, as well as at the roadside, it is an ideal choice where space is limited, especially to contrast with broad crowned trees.

## *New black poplar hybrids*

During the past 20 years substantial numbers of new hybrid black poplars, most of them produced at research stations in western Europe, have been released for general propagation and field cultivation, usually after rigorous testing. Most are selections derived from controlled, artificial crosses between carefully chosen parents of *P. deltoides* and *P. nigra*. Their future role in Britain, whether for timber production or for amenity, can still only be guessed at. But intensive trials on a range of sites in different parts of the country should provide reliable evidence of their potential value within a few years. The best known are listed by country of origin.

### Belgium

*P.* × *euramericana* 'Gaver', 'Ghoy', 'Gibecq', 'Isières', 'Ogy' and 'Primo' have all been well tested in disease trials (see Appendix 2). The results suggest they are satisfactorily resistant to bacterial canker and to *Marssonina brunnea*. Early assessments also showed them to be resistant to *Melampsora* leaf rust fungi. But subsequent observations revealed serious infection of 'Isières' and 'Ogy' by *M. larici-populina* and, as a consequence, the commercial use of these two is now in doubt. A new physiological race of the fungus was thought by Steenackers (1982) to be responsible for the damage. In Belgian trials, all the cultivars have a faster rate of radial growth than 'Robusta' and rotations of about 20 years for the production of veneer logs and sawlogs are already being predicted (Crichton, 1983). In the earliest planted trials 'Ghoy' has proved to be the fastest growing of the six cultivars.

### Germany

A large number of euramerican hybrids have been selected in Germany in recent years and raised to cultivar status. Most of them have been released to horticulture although their cultivation sometimes appears to be quite local. Some of them have been included in field trials in Britain but none has grown especially well. There is little justification for importing other German selections until more is heard of their potential value elsewhere.

Italy

*P.* × *euramericana* 'BL Costanzo', 'Boccalari', 'Branagesi', 'Cappa Bigliona', 'Gattoni', 'San Martino' and 'Triplo' are all fast growing in Italy and it is more than likely that most if not all will prove to be similarly vigorous in other countries that also enjoy long, warm summers. They are considered to be satisfactorily resistant to *Marssonina* leaf diseases which, in Italy, can cause significant loss of increment and wood production. Their potential role in Britain and other relatively cool countries in north-west Europe is questionable. In practice, it is probable that selected products of artificial breeding in Belgium and the Netherlands will be preferred. Attempts to cultivate Italian hybrids in this country in the past have only been wholly successful in the southern half of England. In trials in the north and west, mostly on carefully chosen fertile sites, they have invariably been among the slowest growing cultivars. But even in favourable localities in the south, most have suffered from serious defects. Poor stem and crown form, and susceptibility to bacterial canker have been the most common faults. Since bacterial canker is not present in Italy, new cultivars are not tested for resistance to the disease. Intensive field trials in north–west Europe that include tests to determine proneness to canker, are required before the recent releases can be safely cultivated here.

Netherlands

*P.* × *euramericana* 'Agathe F', 'Dorskamp', 'Flevo', 'Florence Biondi', and 'Spijk' are all growing in stool beds at Alice Holt Lodge and there are semi-mature trees of each in the National Populetum. The accumulated evidence suggests they may not be vigorous enough or sufficiently resistant to bacterial canker to merit cultivation in Britain. A poor stem and crown form on some sites also lessen prospects of planting for wood production.

## Section Tacamahaca – *balsam poplars*

This section is easily the largest of the five and contains about 15 species. Four are native of North America, the remainder are species of Asia. Few of the Asian species are well known outside their natural range and probably less than half has any economic importance. None is likely to be cultivated in Europe other than in parks and gardens, but two or three have considerable potential for breeding. In contrast, two of the North American species produce valuable timber and stands throughout their range have great economic significance. One of them, *P. trichocarpa*, is an important tree in breeding programmes in western Europe where selected cultivars are planted.

### *P. acuminata* Rydb., lance-leaf cottonwood

A native of the dry eastern foothills of the Rocky Mountains from Alberta and Saskatchewan south to Colorado and New Mexico. It arises on the banks of streams and, in the wild, is rarely taller than 15 m. It is sometimes planted as a shade tree in the streets of cities and towns in Colorado and Wyoming. Introduced to Britain in 1916, it was planted to begin with in a few major tree collections but is now rare. A clone obtained by the Forestry Commission in 1939 from the Canadian Forest Service has been so feeble and semipendulous in the nursery and so slow growing in the Populetum that it was not included in field trials. It is doubtful if other provenances would have been more successful.

### *P. angustifolia* James, narrow-leaved cottonwood

Sometimes called willow-leaved poplar, this species is a native of western North America from Saskatchewan and Manitoba south to Nevada, New Mexico and Arizona. It arises on the banks of streams at elevations of 1500 to 3000 m and can reach a height of 15 to 18 m. It is planted as a shade tree in Colorado and Utah. Young foliage has a characteristic fragrant, balsamic odour. A clone imported from Colorado by the Forestry Commission in 1952 is slow growing in the nursery and Populetum, is highly susceptible to *Melampsora* leaf rust and has little to recommend it.

### *P. balsamifera* Duroi non L. (= *P. tacamahaca* Mill.), balsam poplar

Widely distributed in Canada, Alaska and in the United States south to New York, Nebraska, Nevada and Oregon. It can form pure stands or arise in intimate mixture with other pioneer broadleaved trees and some conifers. Mostly occupying valley bottoms, it reaches a height of 30 m. It is an important timber producing tree. It appears to have been introduced to Britain about 300 years ago and, for some time afterwards, was planted in parks, gardens and amenity woodlands largely on account of the strong balsamic odour of the buds, young shoots and foliage. But it has grown less well in this country than most other poplars, and several balsam cultivars introduced since have taken its place. Clones imported from Ontario and Michigan by the Forestry Commission in the 1950s, which it was hoped might be superior to those already in cultivation, have fared indifferently as well. The largest known tree in the British Isles was grown at Faskally House, Pitlochry. It achieved a breast height diameter of 102 cm.

### *P. balsamifera* var. *Michauxii* (Dode) Farwell

A variety occurring wild in north-east North America which has been much confused taxonomically with *P. candicans*. It is of interest to growers in Britain only because it was the male parent of *P.* 'Balsam Spire', a hybrid produced artificially at Harvard University in the mid-1940s with *P. trichocarpa* var. *hastata* as the seed parent. Most of the pleasing characters of the hybrid appear to have been inherited from var. *Michauxii*.

### *P. candicans* Ait., balm of Gilead

Its origin is unknown; whether it is a clone of *P. balsamifera* or a hybrid between *P. balsamifera* and *P. deltoides* is open to doubt. It is quite distinct from *P. balsamifera*, however. Its leaves are broader and usually longer than those of balsam poplar and it always has a clearly defined cordate leaf base (most leaves are obviously heart shaped) and pubescent petioles and young shoots. It has stout spreading branches while *P. balsamifera* has ascending branches. It has been much planted in north-east North America since the mid-18th century, especially in windbreaks and shelterbelts in the Canadian prairies (hence *P. ontariensis* Desf., Ontario poplar). Only a female clone is in cultivation.

Introduced to Britain around 1773, it has been widely planted in parks and gardens for its handsome foliage and intensely balsamic odour. However, it has proved to be extremely susceptible to bacterial canker. Once a specimen is infected, branches soon die back, cankers may arise on the stem and death of the tree can follow surprisingly quickly. The Forestry Commission have long regarded the tree as a major source of the disease and likely to spread infection to neighbouring poplar crops planted for timber production.

*P. candicans* 'Aurora' is a variegated form marketed in Britain for less than 40 years so no large trees are known. Variegated leaves are irregularly discoloured white, cream, pale pink and pale green though, for some time in June and July, all the emerging leaves at the tip of each branch may be entirely white and most of the fully developed leaves dark green. Since variegated leaves are usually most profuse on fast growing shoots, garden trees are often hard pruned during the winter. It is a highly canker susceptible cultivar.

Other variegated forms of *P. candicans* are sometimes encountered. The most handsome has yellow–gold foliage.

### *P. cathayana* Rehd.

A native of north-west China south to Manchuria and Korea, sometimes reaching a height of 30 m. Attempts to cultivate trees in arboreta in this country have not been particularly successful due to damage by frost or bacterial canker or both. Nursery plants are slow growing and stools at Alice Holt Lodge – the first was raised in 1953 from material obtained from Kew

Gardens – produce few cuttings. In favourable seasons large, handsome leaves are found on long shoots.

## *P. ciliata* Wall.

A Himalayan species occurring wild in mixed broadleaved woodland with alder, maple and oak. Veneer logs are sought after for the manufacture of matches. Material from source introduced to Britain during the past 30 years had led to true to name plants being planted in one or two nurseries and arboreta. Clones from Punjab tend to be slow growing at Alice Holt Lodge, however, and attempts to establish trees in the field have not succeeded.

## *P. koreana* Rehd.

A native of Korea north to Primorsk Territory (south-east USSR), sometimes reaching a height of 25 m. A little known tree even within its natural range. Clones imported from the wild by the Forestry Commission in the late 1950s were soon found to be extremely frost tender and susceptible to bacterial canker and all efforts to establish trees in the field were unsuccessful. Although it is a handsome tree with large leaves, dark green above and whitish beneath, specimens are rarely seen in Europe. It was introduced to the United States in 1918 and may have fared better there. A specimen in the Arnold Arboretum, Harvard University, was chosen in the mid-1940s as a male parent to hybridise with *P. trichocarpa*. A clone propagated from one of the seedlings, imported by the Forestry Commission in 1949, frequently grew 3 m in a season in open beds at Alice Holt Lodge. This exceptional vigour in the nursery led to its early inclusion in clonal trials and silvicultural experiments in the field. Regrettably, the hybrid was soon found to be excessively frost tender and prone to bacterial canker. Thus, like all the trees of *P. koreana* raised a decade later from different provenances, most specimens of the hybrid quickly exhibited serious branch and stem dieback soon after planting out. The closely related Japanese balsam poplar, *P. maximowiczii*, may be a more than adequate substitute for this delicate species both for amenity planting and for hybridising with *P. trichocarpa*.

## *P. laurifolia* Ledeb.

A native of Siberia where it is rarely taller than 15 m and appears to have little economic worth. It is best known as the putative (female) parent of *P.* × *berolinensis* (pp.26-27) and other spontaneous hybrids. Introduced to Britain about 150 years ago, it has been described as an elegant tree. In reality, though, it has several defects and understandably, has failed to attract the attention of aboreta curators or local authority park keepers. A female clone known as the volunteer poplar is grown on the Canadian prairies.

## *P. maximowiczii* Henry

A native of north-east Asia and Japan, where it is used in the manufacture of matches. In the wild it can reach a height of 30 m and is one of the largest trees found in the Far East. It is also one of the most ornamental, its leaves, catkins and bark probably being its chief assets. Introduced to Britain around 1913, nearly every effort to establish trees in arboreta and clonal trials has failed, regardless of provenance, because of serious frost injury or attack by bacterial canker. The earliest planted trees may have had their origin in the Soviet Union. Seed and cuttings imported by the Forestry Commission in the 1950s are known to have been collected from trees and stands in Japan and Korea. Of nine clones planted some time later in the National Populetum, only one now survives.

It is a female tree, apparently having considerable potential for amenity planting. Aside from its attractive appearance it has grown soundly for more than 30 years in a location where other *P. maximowiczii* clones and perhaps 15 other poplar species have died or had to be prematurely felled because of disease or winter cold damage. It is one of the earliest trees into leaf – leaves are visible and often half unfolded, like those of *Salix* 'Chrysocoma' (golden weeping willow), before the end of March, and it is one of the few poplars whose foliage turns yellowish-gold, albeit briefly, before falling. It has handsome rather leathery leaves 6–14 cm long, dark green and wrinkled above, whitish beneath, with visible hairs on veins and veinlets on both surfaces, and a characteristic narrow, twisted apex which points downwards and sideways.

Graceful catkins, which occur profusely in all parts of the crown, lengthen throughout the summer becoming 18–25 cm long and only ripen in September or October when large, showy three of four valved capsules begin to open. Its bark, smooth and yellowish to begin with, becomes grey and fissured; for a time the two bark types may be seen on opposite sides of the stem.

*P. maximowiczii* has been hybridised artifically with several other species to produce vigorous progeny characterised by heterosis. The best known were bred in the United States in the 1920s. Ten years later the four fastest growing, hardiest and most disease resistant clones were described by Schreiner and Stout (1934) and given cultivar names. They were registered a few years ago with the International Poplar Commission. They all have the same *P. maximowiczii* seed parent. One, *P.* 'Androscoggin' (page 25), was found in the 1950s and 60s to be the fastest growing tree in Britain. Later, in the 1950s, breeding was undertaken successfully in Poland, again employing *P. maximowiczii* as a female parent, while more recently hybrids have been produced in Belgium, Germany, the Netherlands and Italy using pollen collected in Japan. Now, seed is being collected from 13 provenances in Hokkaido, Northern Honshu and Central Honshu to improve the genetic variety of the species for future selection and hybridisation programmes. The collections are being sponsored by the International Union of Forestry Research Organizations working parties on poplar provenances and breeding. Some of the impetus for this important co-operative venture stems from projects mounted to produce *P. maximowiczii* hybrids with rapid early growth rates for short rotation forestry.

### *P. purdomii* Rehd.

A little-known species of north–west China, probably related to *P. cathayana*. Specimens grown in the south of England have not been authenticated.

### *P. simonii* Carr.

A native of north and central China and Korea. A handsome, medium-sized tree up to 12 m tall with slender branches and a narrow crown. Since about 1960 it has been extensively planted in the north of China to provide protection against the worst of the continental climate. Selection and improvement programmes have been started in the hope of increasing growth rates and yields of wood. Introduced to France in 1862, it was planted for a while in upland areas in western Europe. Now, it is a rare tree even in parks and gardens, having given way to better species and cultivars. It was first planted in Britain at Kew Gardens in 1899; a tree survives but has always grown slowly. At 70 years of age its rate of height increase was found to be only 21 cm per annum. It has hardly ever been planted elsewhere in this country and no other specimens of note are known. Clones tested at Alice Holt Lodge including a Korean provenance have all been highly susceptible to bacterial canker, and none can be recommended for cultivation in Britain.

Two cultivars of *P. simonii* are in commerce but only one has been introduced to Britain. Imported as *P. simonii obtusata* (and sometimes called *obtusata fastigiata*) it is correctly named *P. simonii* 'Fastigiata'. It has upright branches and makes a small, narrow crowned pyramidal tree. The second cultivar, *P. simonii* 'Pendula', has slender branches which arc outwards and become pendulous at their ends.

### *P. suaveolens* Fisch.

A very wide distribution extending from the eastern borders of Turkey through Mongolia and Eastern Siberia to Kamchatka in the north-east of the Soviet Union. It appears to be more closely related to *P. cathayana* than to any other Asian poplar. Over part of its range it is fast growing and wood of economic value is produced. Clones are grown in northern India. As a rule it is rarely taller than 16 m but is an attractive tree and has sweet scented foliage like the balsam poplars of North America. Introduced to Britain in 1834, it appears to have been planted only at Kew Gardens where there were exceptionally small stunted trees. Four clones from different parts of its range in the Soviet Union were imported by the Forestry Commission in 1957 from The Academy of Sciences in Moscow, but all were weak in the nursery and two failed a few years later. Efforts to establish trees in the field were wholly unsuccessful.

*P. suaveolens* 'Pyramidalis' a narrow-crowned, fastigiate tree, has been in commerce in the United States since 1928 but has not been imported to Britain.

## *P. szechuanica* Schneid.

A little known species of western China and of limited economic value. It reaches a large size, often exceeding heights of 30 m, and makes a handsome tree with striking buds and foliage. The buds are purplish and viscid, and most leaves are reddish when young, becoming bright green above, whitish beneath, and reaching a length of 10 to 20 cm. Leaves on nursery plants are especially colourful. Introduced to Britain in 1908, only three notable specimens now survive, two at Stourhead, Wiltshire, the other at Westonbirt Arboretum, Gloucestershire. All are taller than 15 m. A clone tested by the Forestry Commission in the 1950s was moderately susceptible to bacterial canker and extremely frost tender. To establish trees satisfactorily, shelter from late spring frosts appears to be essential. If suitable sites can be found, it is a tree well worth cultivating.

## *P. trichocarpa* Torr. and Gray, western balsam poplar (black cottonwood)

A very wide distribution from southern Alaska south through western Oregon to the islands of western California and the southern slopes of San Bernandino Mountains; eastward through British Columbia to the valley of the Columbia River and to the plains of Montana and Idaho. It is easily the largest broadleaved tree of western North America and reaches its greatest size near sea level in all the coastal regions north of California where it can be 60 m tall. Away from the sea and southward it may reach a height of 30 to 40 m. It ascends to elevations of 1800 m on the western slopes of the Sierra Nevada. Dense pure stands may arise in valley bottoms but in the northern parts of its range it more often occurs in mixture with Sitka spruce, Douglas fir and western red cedar. Throughout its range it produces valuable timber and most stands have considerable economic worth. Its botanical characteristics and behaviour vary appreciably from one part of its range to another; where it overlaps with *P. balsamifera* in the north the two can be difficult to tell apart and natural hybrids may arise.

Introduced to Britain in 1892, it was soon recognised as the fastest growing of the balsam species and one that could be cultivated in woodland at close spacing and on more difficult lowland sites. It proved to be particularly tolerant of conditions in the north and west of the country, where the cooler and wetter summers, and rather more acid soils than are encountered in the south, seriously retard the growth of black poplar hybrids.

There is some evidence to suggest that the first few clones of *P. trichocarpa* introduced to Britain had a northern, probably Canadian, origin. These turned out to be susceptible to bacterial canker, so limiting their commercial usefulness. However, in the 1950s, 16 newly selected clones were imported by the Forestry Commission from North America for inclusion in disease and field trials. Although most of them were chosen in woodlands in Canada, those selected in stands in the United States proved to be of particular interest. The best were from Washington State or close to its borders. In the 1970s, two of the most disease resistant and vigorous clones imported from Washington State were released to the nursery trade in this country and approved by the Forestry Commission for planting for timber production. They were named 'Fritzi Pauley' and 'Scott Pauley'. At about the same time the first was released in Belgium, France, Germany and the Netherlands, where a full account was published, and the second was released in Germany, where a botanical description was prepared. In 1973 seed of 17 provenances were also imported. Nine seed lots were collected in Washington State, five were collected farther south in Oregon and three were collected in Canada. In 1976 rooted cuttings raised from the progeny were planted in experiments at Alice Holt and Cannock, Staffordshire.

## *P. trichocarpa* var. *hastata* Henry

A variety distributed in north California and northward, and into the drier interior regions where it is smaller than the oceanic type *P. trichocarpa*. It has large leaves, longer than broad, in comparison with the Californian form which has comparatively small leaves often broader than long. A tree growing in the Arnold Arboretum, Jamaica Plain, Massachusetts, USA, was the female parent of *P.* 'Balsam Spire' (pp. 25-26).

## *P. trichocarpa* 'Fritzi Pauley'

A female cultivar with very high resistance to bacterial canker and *Marssonina* leaf disease, and with satisfactory resistance to *Melampsora* leaf rusts. It grows rapidly on a wide range of soils including heavy clays; at best its volume production is 170 per cent greater than that of 'Robusta'. Usually straight stemmed, it has fine, easily pruned branches though heavy pruning may lead to growth of epicormics. It can be cultivated in woodland and at close spacing for pulpwood and other small sized produce. Its only defect is a proneness to stem fracture, usually in the upper crown, in high gusty winds. Although damaged trees recover quickly, some care is clearly needed in siting specimens in places accessible to the public. Selected near Mount Baker in Washington State, it was introduced to Britain in 1950 and, at about the same time, to other countries in western Europe. It has been used as a seed parent in breeding programmes in Belgium, especially in crossings with *P. deltoides*. It is approved by the Forestry Commission for planting for wood production.

## *P. trichocarpa* 'Scott Pauley'

A vigorous cultivar from Washington State only slightly less resistant to bacterial canker than 'Fritzi Pauley' and as resistant as this to foliar diseases. It has a very fine form and is not prone to grow epicormics unless very heavily pruned. It is not damaged by wind. Approved by the Forestry Commission for wood production, it appears to be a better prospect than 'Fritzi Pauley'.

Other clones of *P. trichocarpa* are under trial in this country. 'Blom' and 'Heimburger' released in the Netherlands are probably slower growing than 'Fritzi Pauley' and 'Scott Pauley'; but 'Columbia River' and 'Trichobel', artificial hybrids between two selected clones of *P. trichocarpa*, are probably more vigorous and probably merit extended trial in different parts of Britain. They were selected and tested in Belgium (see Appendix 2). A clone called 'Muhle Larsen', released in Germany, may be too susceptible to bacterial canker to be cultivated here but a synthetic, poly-clonal cultivar named 'Bruhl', at present under trial in that country, may be suitable. It is a mixture of eight clones of *P. trichocarpa* selected in Oregon.

*P. trichocarpa* has been much used in artificial breeding programmes in several European countries as well as in the United States. Until recently only a comparatively small number of clones was available for crossing with other species. Since the 1960s, however, the selection of many more clones in the wild and the collection of seed throughout the species range has greatly increased the genetical variety of parents available to tree breeders. The best known *P. trichocarpa* hybrids in Britain are *P.* × *generosa* (page 28), *P.* 'Androscoggin' (page 25) and *P.* 'Balsam Spire' (pp. 25-26). Though no longer planted on account of its extreme susceptibility to bacterial canker, *P.* × *generosa* remains pre-eminent amongst poplar hybrids. It was the first to be bred artificially.

## **P. tristis** Fisch.

A small, little known tree of central Asia. It has been cultivated for amenity in the north-west of the Soviet Union where some provenances are so winter hardy they can be grown north of the Arctic Circle. Probably introduced to Britain in 1896, the first tree, planted at Kew Gardens, survived for only a few seasons. Efforts to establish it elsewhere have been unsuccessful and none of the specimens located outside arboreta have been verified botanically. Stools at Alice Holt Lodge have remained alive for only a few years. Two trees planted recently at Kew are growing tolerably well but so far have a lax habit. They are very early into leaf. Introduced to North America in the early part of the 19th century, it has been cultivated mainly on the Canadian prairies where clones have been selected for planting in shelterbelts.

## *P. yunnanensis* Dode

A native of Yunnan, south-east China. A decorative tree with handsome leaves up to 15 cm long, bright green above, whitish beneath with a red mid-rib and petiole. It is the most southerly of the balsam poplars and in mild, frost free localities it can retain its foliage well into winter. With only odd exceptions, specimens have proved difficult to establish in Britain because of serious frost injury, though trees planted by the Forestry Commission in the 1950s were badly damaged by bacterial canker after only a few years in the field. There is slight dieback of young shoots on two recently planted trees at Kew, which are probably the last balsam poplars into leaf. Stools at Alice Holt Lodge, producing highly ornamental foliage, have tended to have a short life.

## Hybrid balsam poplars
### *P.* 'Androscoggin'

A male hybrid bred artificially in 1924 for the Oxford Paper Company at Rumford, Maine, USA. The seed parent was *P. maximowiczii*, the pollen parent *P. trichocarpa*. Both were selected in the New York Botanical Garden, Bronx Park, New York City. 'Androscoggin' closely resembles *P. maximowiczii* in appearance and botanical characteristics. It was raised in a nursery at Frye, Maine, with about 13 000 other hybrid seedlings obtained from 99 different cross combinations between 34 species and cultivars. Ten years later it was described and named along with nine other hybrid clones (Schreiner and Stout, 1934). Imported to Britain by the Forestry Commission in 1937, it was widely planted in trials in the 1950s, until outbreaks of bacterial canker under natural conditions led to a halt in the planting programme. The most serious damage occurred at a site in west Scotland were several young trees died only 2 years after becoming infected. In the majority of plantings it was invariably among the fastest growing clones; in some trials it was easily the most vigorous. In the 1950s and 60s, specimens at Quantock Forest, Somerset, were considered to be the fastest growing trees in Britain. When felled at 35 years of age, the largest tree was 34 m tall and had a breast height diameter of nearly 60 cm. In almost all trials trees were straight stemmed and, since branches tended to arise together in annual whorls, much of the upper stem between whorls of branches was usually suitable for rotary peeling. High pruning was easily carried out using saws on extendable rods. Clear white, flawless veneers and splints of moderate strength were produced during rotary peeling tests undertaken by the match industry on billets cut from pruned logs.

Two other *P. maximowiczii* × *trichocarpa* hybrids raised by Stout and Schreiner and imported in 1948 ('NE 42') and 1954 ('NE 388') also proved to be too susceptible to bacterial canker to be released for general cultivation, as was a clone of the same parentage imported from Poland in 1959. Prospects of producing fresh hybrids with the desirable qualities of 'Androscoggin', but showing resistance to bacterial canker, are not unrealistic, however, as seed and cuttings of *P. maximowiczii* from provenances in Japan are becoming available to tree breeders in Europe.

### *P.* 'Balsam Spire'

A female hybrid bred artificially at Harvard University from *P. trichocarpa* var. *hastata* (female parent) and *P. balsamifera* var. *Michauxii*. Imported by the Forestry Commission in 1948 along with five full sib clones, it was planted in 15 trials spread across the country during the next 10 years. In 1957, as a result of its satisfactory behaviour in the trials, it was released to the private forestry sector and horticulture and added to the list of cultivars eligible for Forestry Commission planting grants.

To begin with, it appeared in nursery stock lists as *P. tacamahaca* × *trichocarpa* 32. Hardly surprisingly, nurserymen soon objected to this cumbersome name and a shortened version – *P.* 'tacatricho 32' – quickly appeared in catalogues. Almost inevitably this was further abbreviated to 'TT32'. Understandably, poplar research workers concerned with the registration of cultivar names with the International Poplar Commission objected to this abbreviated nomenclature and the Forestry Commission were asked to submit an acceptable cultivar name for the hybrid. In 1982 'Balsam Spire' was put forward and approved.

Like *P. trichocarpa* and some other *P. trichocarpa* hybrids, 'Balsam Spire' grows well on a wide range of sites and is seemingly well suited to heavy clay soils, comparatively dry and acid sandy loams and to damp, somewhat badly drained soils. It is a useful tree for the cooler and wetter parts of the country. Its rate of height growth is usually rapid, at best exceeding 1.5 m per annum for the first 15 years. It easily keeps pace with other vigorous balsam poplars and, other than in exceptional seasons, betters all black poplar hybrids. On fertile sites in the south, trees can reach 10 m in 6 years and 20 m in 13 years. Specimens which remain fast growing can be over 30 m tall after only 30 seasons. The tallest measured tree had a height of 33 m at 33 years of age. Unfortunately, its rate of radical increment tends to be lower than that of other poplars displaying vigorous height growth. Even open-grown trees may be thinner stemmed, age for age, than specimens of other clones. None the less, trees large enough for rotary peeling can be produced on a wide range of sites in 30-35 years. Its comparatively slow radial growth seems likely to be due to a low leaf area index, directly related to fine and short primary branching and fine, sparse secondary branching.

'Balsam Spire' has the narrowest crown of any commercial poplar. It has a fastigiate habit, nearly all branches arising at an angle of 45° to the stem, and a cylindrical, rarely forking stem. Its branches are fine and short. Partly on account of its narrow crown it can be cultivated at narrow spacing to produce pulpwood on short rotations. Cultivation has been successful at spacings as close as 2 × 2 m.

It is one of the earliest trees into leaf and is much used to shelter fruit and farm crops. Flushing is early enough to protect most early flowering fruit varieties. Planting in single row windbreaks is favoured with the trees only 1.5 m apart. Its narrow crown never encroaches onto cropped land; in any case crown development can be restrained by mechanical trimming. It is free of suckers but because it has wide spreading roots, which compete for moisture and nutrients and may hinder soil cultivation, it is best used in perimeter windbreaks. Female catkins dropping in the spring can be a nuisance since the visual appearance and value of certain crops can be affected by deposits of the white down. A tendency to lose its leaves in late summer is a minor disadvantage since maximum protection may not be given to top fruit at harvest time.

In bacterial canker trials 'Balsam Spire' displayed moderate susceptibility on inoculation. But in practice in conditions leading to natural infection damage is usually limited to small twig cankers. Branch cankers reducing growth rate and stem cankers lessening timber value are rare. On low lying sites liable to winter and spring frosts, heart shake or frost crack can be a serious problem. 'Balsam Spire' is resistant to *Marssonina* and *Melampsora* leaf fungi.

## Hybrids between sections Tacamahaca *and* Leuce

### *P. alba* × *P. trichocarpa*

A hybrid first bred in Germany more than 30 years ago and later reproduced in Canada. The only clone surviving in the Forestry Commission poplar collection was raised in Stuttgart in 1954. It was imported from Canada in 1959. Stools have since survived in nurseries at Alice Holt Lodge but annual shoot growth is nearly always weak and hardwood cuttings suitable for propagation are rarely produced. Though rooted stocks can be raised from softwood cuttings in mist, they also grow feebly. Attempts to establish trees in the field have been unsuccessful. Juvenile leaves on plants in the nursery are most like those of *P. alba* stocks. They are lobed, each lobe with a few teeth, and the undersurface is always covered with a greyish-white felt. They are invariably longer than broad. So far as is known, all the hybrids including those bred in Canada have had *P. alba* seed parents and pollen appears to have been collected only from *P. trichocarpa* clones commonly grown in Europe.

## Hybrids between sections Tacamahaca *and* Aigeiros

### *P. laurifolia* × *P. nigra*

*P.* × *berolinensis* Dippel, Berlin poplar
A natural hybrid judged to have arisen from *P. laurifolia* (female parent) and *P. nigra* 'Italica', Lombardy poplar. Both sexes exist. A

female clone, believed to be the first cross, arose in the Botanic Garden of Berlin around 1865. It appears to have been little planted. In contrast, a male clone, thought to have originated some time later in France, has been widely planted in Europe and North America. At first it was marketed as *P. certinensis*. The two forms are similar in crown and stem habit, leaf size and shape and rate of growth. The male clone is considered to be very hardy, withstanding long, cold winters and hot, dry summers. Consequently it was much planted for a time in central and western Europe and on the Canadian prairies. Its ascending branches and columnar crown encouraged its use in single row windbreaks. In recent years its susceptibility to foliar diseases, causing premature leaf fall and a reduction in growth rate, together with the increasing availability of faster growing and disease resistant cultivars has greatly lessened its use both for wood production and for screening.

Two male trees were planted at Kew Gardens in 1889 and both grew sufficiently well for some years to encourage planting in parks and arboreta in other parts of the country. Later, in the 1930s, it was sometimes included in planting schemes at the roadside though, perhaps surprisingly, it was rarely planted for wood production. In the 1950s, however, it was found to be resistant to bacterial canker and, for a short while, it was added to the list of cultivars eligible for Forestry Commission planting grants. But in several field trials it proved to have such a low rate of radial growth, perhaps due to habitual premature defoliation, that steps to make cuttings and plants available to nurserymen and timber growers were halted. There has been little planting since and *P. × berolinensis* is now a rare tree. Variants have been reported in Britain and Germany, some with comparatively wide crowns. A clone reported by the Forestry Commission in 1957 from Kursk Province in the Soviet Union has atypical foliage but is authentic in other respects.

## *P.* 'Frye', *P.* 'Rumford' and *P.* 'Strathglass'

Full sibs bred in 1924 for the Oxford Paper Company, Maine, USA, from *P. nigra* (seed parent) and *P. laurifolia*. They were selected from a total of 377 seedlings. Ten years later they were described and given cultivar names along with seven other hybrids (Schreiner and Stout, 1934). They were introduced to Britain by the Forestry Commission in 1937. 'Rumford' and 'Strathglass' were found to have a slow rate of radial growth and to be subject to leaf diseases and, though moderately resistant to bacterial canker, planting was limited to a few field trials and arboreta. Most of the earliest planted trees have since been felled and specimens of the two cultivars are rare. The few still standing have sparse, open crowns and there is little to commend them. In contrast 'Frye' has proved to be reasonably vigorous, on better sites reaching 18 m in 20 years, and handsome, shapely trees have usually been produced. Most have developed a healthy, dense crown free of dieback. Unfortunately, specimens on open ground tend to lean with the prevailing wind and epicormic shoots develop after pruning. A specimen in the populetum was 26 m tall at 33 years of age. 'Frye' and 'Strathglass' are female trees; 'Rumford' is male.

## *P. × petrowskyana* Schneid.

A putative hybrid believed to have arisen in the Soviet Union around 1880. Its parentage is unknown, but it is botanically similar to *P. × berolinensis* and may, therefore, be a *P. laurifolia × P. nigra* 'Italica' hybrid. Three clones in the possession of the Forestry Commission display marked dissimilarities in resistance to disease and foliar characteristics, and almost certainly have different origins. Their nomenclature is questionable. The disparities have been most marked in the National Populetum, where specimens of a clone obtained in 1953 from Kew Gardens were almost killed by continuing bacterial canker infection and had to be felled, while an adjacent tree, grown from a clone obtained from Ryston Hall, Downham Market, Norfolk, has remained healthy and vigorous since planting. It has an attractive, dense crown and columnar habit. They are all female trees.

## *P. × rasumowskyana* Schneid.

Considered to be a hybrid and, like *P. × petrowskyana*, may also be a cross between *P.*

*laurifolia* and *P. nigra* 'Italica'. It is thought to have originated in the Soviet Union around 1880. Three clones in the possession of the Forestry Commission are closely allied to *P. × berolinensis*, though one has atypical, attractive pink veins and a second is more susceptible to bacterial canker than the other two. The nomenclature of all is doubtful. They are male trees.

## *P. × octorasdos* and *P. × wobstii*

These are also near to Berlin poplar botanically and may have the same parents. At any rate they are probably best grouped with *P. × berolinensis*, though both require botanical elucidation. Thirty-year-old specimens in the National Populetum are slow growing and have thin, open crowns showing much dieback. They are male trees.

## *P. maximowiczii* × *P. nigra*

*P.* 'Rochester'. A female hybrid bred for the Oxford Paper Company, Maine, USA, in 1924 from *P. maximowiczii* (female parent) and *P. nigra* 'Plantierensis'. Selected from 145 seedlings, it was described and named ten years later (Schreiner and Stout, 1934). Imported by the Forestry Commission in 1937, it was found to be reasonably vigorous in full trials for the first ten years or so but growing less well later and to be increasingly prone to forking and heavy branch development. It shows no outward signs of *P. nigra* parentage and could pass as pure *P. maximowiczii*. Though officially released to the nursery trade in the Netherlands in 1972, it is too susceptible to bacterial canker to be put forward for general planting in Britain.

## *P. trichocarpa* × *P. deltoides*

*P. × generosa* Henry

The first artificially bred hybrid between a balsam poplar and a black poplar, produced at Kew Gardens in 1912 from *P. deltoides* 'Co-data' (seed parent) and *P. trichocarpa*. Several seedlings were raised and propagated. The cross was repeated two years later. The progeny proved to be vigorous and for perhaps three decades *P. × generosa* was widely planted in Britain for amenity, often at the roadside and in parks and gardens in preference to other poplars. It was soon found to be extremely susceptible to bacterial canker and to *Melampsora* leaf rust fungi, however, and interest in its use subsided just as quickly. It is rarely planted nowadays and mature specimens are few and far between. More often than not, *P. × generosa* was usually the first cultivar to be attacked by bacterial canker in mixed plantings and, until felled, affected trees remained a source of infection and a focus from which the disease was able to spread to other plantings. Its disappearance from the countryside has been a boon to modern poplar growers. A proneness to stem breakage also constituted to a decline in its rate of planting. The number of clones in cultivation, and the extent of their variation in appearance and behaviour, is unknown. Both male and female of this historically important hybrid were planted at one time.

Modern hybrids between the two North American species *P. trichocarpa* and *P. deltoides* are probably best grouped under the collective name *P. × interamericana* Van Broekhuizen (c.f. *P. × euramericana* for hybrids derived from *P. deltoides* and *P. nigra*). This collective epithet including a Latin diagnosis was first used by Broekhuizen (1972) in his descriptions of three new cultivars bred artificially at the Dorschkamp Research Institute for Forestry and Landscape Planning, Wageningen, The Netherlands, and released to the nursery trade in that country in the spring of 1972. A history of the three cultivars, namely 'Barn', 'Donk' and 'Rap', the results of tests carried out to assess their resistance to bacterial canker, *Melampsora* and *Marssonina*, and a brief summary of their expected uses were prepared by Koster (1972). The clones were imported by the Forestry Commission in 1974, planted in the Populetum at Alice Holt Lodge in 1976 and included a field trial at Oxted, Surrey, in 1979.

## *P.* 'Barn' and 'Donk'

From a female *P. deltoides* fertilised by *P. trichocarpa*. They are fast growing clones but are too susceptible to bacterial canker to warrant their release in Britain for general planting.

## P. 'Rap'

Raised from a female *P. trichocarpa* pollinated by *P. deltoides*, it has proved to be a fast growing cultivar matching the best *P. trichocarpa* selections in height increment. Because of the fast rate of growth of shoots on nursery stools cut-back annually – shoots 4 m long are sometimes produced in a single season – 'Rap' was included in a series of Forestry Commission biomass experiments started in 1981. Fields of trees coppiced at the end of the first growing season at spacings of 1 × 1 m (10 000 stools per ha) and 2 × 2 m (2500 stools per ha) are being compared on two and four year rotations. Preliminary results show that the dry matter production of 'Rap' compares favourably with vigorous willows and is consistently better than that of alder, eucalyptus and southern beech. Unfortunately, since the start of the biomass study, 'Rap' has been found to be excessively susceptible to bacterial canker and will not be included in any further field experiments. It is far too prone to the disease to be recommended for general use. Erumpent cankers have developed on the stems of specimens in the Populetum.

Of far greater potential value to growers in Britain are various hybrids bred at the Government Poplar Research Station, Geraardsbergen, Belgium (see Appendix 2). The cultivars were selected because of their extremely fast growth rate in trials planted in the 1970s, their fine stem and crown habit and for showing resistance to bacterial canker, *Melampsora larici-populina* and *Marssonina brunnea*, in rigorous tests. Named 'Beaupré' and 'Boelare', the cultivars are full sibs derived from *P. trichocarpa* 'Fritzi Pauley' pollinated by an intraspecific *P. deltoides* hybrid bred from Iowa and Missouri provenances. In the Belgian trials, both cultivars have grown vigorously since planting, faster than any selected *P. trichocarpa* cultivar and it is already being forecast that their use in commercial plantations will reduce rotations to 20 years or less. In some trials in which trees are approaching veneer log dimensions (breast height diameter = 45 cm), their volumes are up to 200 per cent greater than the volumes of 'Robusta' controls of the same age. They are rightly regarded as the fastest growing broadleaved trees in western Europe.

All types of cuttings and sets of these and other *P.* × *interamericana* cultivars, as well as most selected clones of *P. trichocarpa*, have high rates of root initiation and development. As a consequence, rooted plants are readily propagated from hardwood cuttings in open nursery beds and, in Belgium, sets are being employed increasingly as planting stock. In trials established by the Government Poplar Research Station 2-year sets are used. They are often 3 to 4 m tall. Survival rates are rarely less than 100 per cent.

The availability of several full sib cultivars in Belgium is prompting interest in the adoption of poly-clonal mixtures in plantations. Since the cultivars come into leaf at the same time, display similar growth patterns during the summer and have almost identical growth rates and stem and crown habits, both line and individual tree mixtures are culturally acceptable. Such mixtures clearly spread the risk of losses due to biotic and climatic agencies.

Tests carried out on wood samples of the new cultivars are encouraging. They had comparatively high basic densities and peeled satisfactorily to produce fine, smooth veneers. In this connection, it appears probable that the cultivars are not inclined to develop epicormic shoots, a habit which degrades the wood quality of so many poplar species and hybrids.

A programme of back-crossing is also being undertaken at the Government Poplar Research Station in Belgium. The results, to say the least, are extremely promising. Where *P. trichocarpa* × *P. deltoides* hybrids have been back-crossed with *P. deltoides*, the progeny tend to have the highest general resistance to poplar diseases and produce wood with higher basic densities than any other group of hybrids. On the other hand, progeny derived from back-crossing with *P. trichocarpa* are usually faster growing and selected clones may be more tolerant of difficult site conditions. 'Beaupré' and 'Boelare' are important parents in the programme. Some of the back-crosses are very complex, being derived from up to five clones of very differing provenances.

The promise of shorter rotations, high disease resistance, improved yields and a greater site tolerance offered by the new *P.* × *interamericana* cultivars should not be overlooked in Britain. Every encouragement should be given to extend the planting of these cultivars once they have been approved for use in this country.

## *P. trichocarpa* × *P. nigra*
### *P.* 'Andover'

A male hybrid bred for the Oxford Paper Company, Maine, USA in 1924 from *P. nigra* var. *betulifolia* (female parent) and *P. trichocarpa*. Selected from 209 seedlings, it was described and given cultivar status ten years later (Schreiner and Stout, 1934). It was imported by the Forestry Commission in 1937 and included in only two field trials. It proved to be too slow growing and too susceptible to *Melampsora* leaf rust and bacterial canker to merit further planting. High pruning may lead to the profuse development of epicormic shoots.

### *P.* 'Roxbury'

A male hybrid bred by Stout and Schreiner from *P. nigra* (seed parent) and *P. trichocarpa*. It was selected from 200 seedlings. Imported in 1937, it was included in five Forestry Commission trials but was not sufficiently vigorous or attractive enough to warrant approval for general use. Though undoubtedly resistant to bacterial canker, branches are inclined to die back, and crowns become increasingly open and thin. Epicormic shoots invariably arise after high pruning. Trees are likely to lean with the prevailing wind on exposed sites.

*P. trichocarpa* × *P. nigra* hybrids have also been bred artificially in Canada and, latterly, in Belgium. Clones imported from Canada in 1949 by the Forestry Commission all have a tendency to grow epicormic shoots, sometimes in surprisingly large numbers, and are effectively debarred from timber production on this account. Even a light pruning leads to their development. This is unfortunate as one of the clones grows well in Britain and is only slightly susceptible to *Melampsora* leaf rust and bacterial canker. It is satisfactorily resistant to *Marssonina*. A specimen in the poplar collection at Alice Holt Lodge was 28 m tall at 33 years of age. None of the Belgian hybrids has been imported.

## *P. candicans* × *P.* × *berolinensis*
### *P.* 'Geneva'

A female tree bred for the Oxford Paper Company, Maine, USA, in 1924 from *P. maximowiczii* (seed parent) and *P.* × *berolinensis*. It was selected from a total of 112 seedlings and described and named ten years later (Schreiner and Stout, 1934). Imported by the Forestry Commission in 1937, it was found to be moderately resistant to bacterial canker but never grew fast enough in early trials to warrant its approval for timber production. This is perhaps fortunate, as older trees have a tendency to die back so that, as a consequence, crowns become progressively thinner and growth rates are reduced even further. In 1966 it was released to the nursery trade in the Netherlands on account of its reasonable growth on exposed sites and satisfactory resistance to *Marssonina* and *Melampsora* leaf diseases.

### *P.* 'Oxford'

This has the same parents and history as 'Geneva'. Imported by the Forestry Commission in 1937 and planted in eight field trials. It proved to be fast growing in most trials, usually increasing in height by more than 1 m per year, and was invariably straight stemmed and free of epicormic shoots. The finest trees were grown in Quantock Forest, Somerset, where at 11 years of age two specimens were felled to provide billets large enough for rotary peeling by the match industry. They were peeled, along with other cultivars, to compare veneer quality and match splint strength. The trees remaining in the plot had a mean height of 23 m and a mean diameter at breast height of 35 cm at 18 years. Unfortunately, bacterial canker spread into both young and old plots and, at a few sites, caused serious damage, and though 'Oxford' had shown moderate resistance to the disease in inoculation tests, the outbreaks effectively precluded its release for timber production. In canker free localities, however, it remains attractive and can develop into a fine specimen tree. Its stem and crown habit are good, and it has handsome foliage virtually indistinguishable from that of *P. maximowiczii*, its female parent, and a pleasing white bark like this species. It was officially released to the nursery trade in the Netherlands in 1966 because of its resistance to *Marssonina* and *Melampsora* leaf disease.

## *Section* Leucoides

### *P. heterophylla* L., swamp cottonwood

This has a wide distribution in the eastern United States overlapping with that of *P. deltoides* in Georgia and South Carolina. It extends from Connecticut in the north to Georgia in the south and as far west as Illinois, Missouri and Louisiana. Though it can reach a large size, up to 30 m tall, it has no commercial value and is rarely cultivated, even in arboreta. It has large leaves, sometimes 18 to 20 cm long and nearly as broad, densely pubescent white to begin with, becoming smooth and green later. The leaves are extremely uniform in shape and it is probably this marked change in hairiness during the season which accounts for its specific name. Introduced to Britain in 1765, it has proved impossible to establish in this country; very few trees have survived for more than a few seasons and none appears to have exceeded a height of 3 or 4 m. Seed imported by the Forestry Commission in 1958 from Union County, Illinois, germinated well but none of the trees was vigorous and they lived for only a few years. Stools in nurseries at Alice Holt Lodge have soon failed. Swamp cottonwood is a handsome tree in the nursery and in a healthy state in the field, and it is a pity that provenances suitable for cultivation in Britain have not been found. Like other *Leucoides* poplars, it is impossible to propagate from hardwood cuttings in open nursery beds.

### *P. lasiocarpa* Oliv., Chinese necklace poplar

This is widely distributed in central and western China, where it can reach a large size and has some economic importance. It is perhaps the most interesting poplar botanically, since trees may be monoecious with both unisexual male and female flowers arising in the same polygamous catkins. The flowers are self-pollinating. It also has the largest and most decorative leaves of any popular species or hybrid. They reach their greatest size on vigorous nursery stocks and can be up to 35 cm long and 23 cm wide. Yet even mature trees in the field may have leaves 30 cm long on fast growing shoots, while they are hardly ever less than 15 cm long on dwarf shoots. They are leathery, pubescent beneath and the base is always cordate. The rich red main veins, midrib and petiole, which can be about half as long as the leaf blade, add to the attractiveness of the foliage. Imported to Britain in 1900, it has been little planted and specimens are far and few between. The largest tree in the Botanic Garden, Bath, Avon, is 27 m tall and still growing strongly. There are two rather smaller trees at Westonbirt Arboretum, Gloucestershire. Propagation difficulties may account for its rarity. Like other *Leucoides* poplars it cannot be raised from hardwood cuttings in open beds and, although a monoecious clone is cultivated in this country, no systematic seed collections are made. It is normally raised by grafting and can be successfully grafted on to rootstocks of a wide range of cultivars. It can also be propagated from softwood cuttings under mist as well as from truncheons (small branches) in the field. It is a pity that such a remarkable and handsome tree is not generally available in horticulture.

### *P. violascens* Dode

On general appearance this species, a native of China, is associated with the section *Leucoides*, where it is placed by most authorities. The classification is sometimes questioned, however, and it is assigned to the section *Tacamahaca*. Dode's description is insufficiently precise for a judgement to be made. It is similar botanically to *P. lasiocarpa*, though and appears to be related to this species more than any other. It has large leaves, up to 15 cm long, subcordate at the base and covered with long soft hairs beneath. Young leaves have a bright purple undersurface.

It is a rare tree in this country and perhaps it is too frost tender to be cultivated successfully. Trees raised by the Forestry Commission from cuttings imported from the Netherlands in 1952 were severely injured in the populetum during their first winter. They were seriously damaged by bacterial canker in the following growing season and had to be felled. The clone was raised from hardwood cuttings, possibly casting doubt on its association with the section *Leucoides*.

### *P. wilsonii* Schneid.

A native of central and western China where it can reach a height of 25 m. It is similar to though less striking than *P. lasiocarpa*. It has smaller leaves, usually 8 to 18 cm long and rarely longer than 23 cm even on vigorous plants in the nursery. The leaves are soon glabrous on both surfaces, though for a short time when unfolding they are pubescent beneath. Young leaves briefly have a reddish undersurface but never have red veins, midribs or petiole like *P. lasiocarpa*. Imported to Britain in 1907, it has been little planted and specimens are rare. It is impossible to propagate from hardwood cuttings in open beds, which partly accounts for its infrequency, but can be raised by grafting. Softwood cuttings root readily under mist. Two clones obtained by the Forestry Commission in 1952 and 1953 soon failed after planting in the populetum, and shoots of both have since died in the nursery. The largest trees in the British Isles are in County Cork and County Meath, Ireland. Only female trees appear to be in cultivation.

*Plate 1.* On moist fertile sites poplar species will produce an early timber return. (*38491*)

*Plate 2.* 20-year-old line-planted P. 'Robusta', a black poplar hybrid. (*3914*)

*Plate 3.* Black poplar hybrids P. × *euramericana* in winter: P. 'Heidemij', P. 'Eugenei' and P. 'Gelrica'. (*28586*)

**Plate 5.** Lombardy poplar *P. nigra* 'Italica' is the best known and most commonly planted fastigiate tree. (*13748*)

**Plate 4.**
When open-grown, grey poplar *P. × canescens* forms an attractive specimen tree. (*36485*)

**Plate 6.**
3-year-old poplars with alternate bays of fallow and wheat intercropping. Herefordshire, *c.* 1975. (*A. Beaton*)

*Plate 7.* Leaves of *P.* 'Heidemij' showing symptoms of infection by poplar mosaic virus. (*15989*)

*Plate 8.* Bacterial canker of poplar caused by *Xanthomonas populi*. (*38155*)

*Plate 9.* Larvae of the large red poplar leaf beetle *Chrysomela populi* and resultant feeding damage on a poplar leaf. (*4307*)

**Plate 10.** P. 'Beaupré', a new P. × *interamericana* clone resulting from artificial hybridisation of two North American species P. *trichocarpa* and P. *deltoides*; 3-year-old trees in a clonal trial (planted 1987) at Bedgebury, Kent. (*C. J. Potter*)

**Plate 11.** P. 'Beaupré', 10 years old, in a trial plantation in Belgium, underplanted with alder coppice. (*C. J. Potter*)

**Plate 12.** P. 'Beaupré', 17 years old, in a trial plantation in Belgium. (*C. J. Potter*)

**Plate 13.** P. 'Primo', 17 years old, in a trial plantation in Belgium. This is a new P. × *euramericana* clone, a black poplar hybrid between the species P. *deltoides* and P. *nigra*. (*C. J. Potter*)

# Chapter 3
# Field Recognition of the Chief Poplars Grown in Britain

## Botanical key

Many woodland and ornamental trees are readily identified from foliage specimens alone. With the possible exception of two or three taxa this is not so with poplars. This is because of the wide variation in the size and shape of leaves on the same tree, the apparent similarity of leaves from different trees and the employment of variable and overlapping characters in botanical descriptions and analytical keys. As a consequence the leaves selected for examination may be imperfect or even atypical and it may not be possible, after sifting botanical evidence, to reach a conclusion confidently. The only system permitting reliable poplar identification from foliage alone was devised by Broekhuizen (1964). However, it was designed only to differentiate commercial cultivars in nursery beds, and while the morphological principles may be applied also to leaves on vigorous epicormic shoots the method has not been adapted to identify trees in the field. Since absolute measurements are carried out on a statistical sample of carefully selected leaves, Broekhuizen's classification is highly reliable and the procedure can be easily extended to include newly released cultivars.

Conventional accounts of poplar species and cultivars normally include details of branchlets, buds, flowers and fruit as well as descriptions of leaves. Bean (1976) provides comprehensive narratives of nearly 50 major taxa and Rehder (1962) separates 31 taxa in an analytical key employing foliar, bud, flower, fruit and branchlet features. In contrast Mitchell (1974) classifies 13 common species and hybrids using only crown and branch shape and bark colours.

In the key that follows on pages 34 to 36, customary morphological minutiae have been combined with characteristic crown and branch form to effect the separation of taxa. Two further features of great assistance in differentiating some cultivars have been included in the key to increase its reliability. These are the sex of the tree and the relative date of flushing in the spring. These distinctive peculiarities are considered in detail after the key.

Difficult botanical terms have been avoided in the key. It is impossible, however, to describe leaf-shapes or leaf-bases without the use of some technical expressions. So that these can be

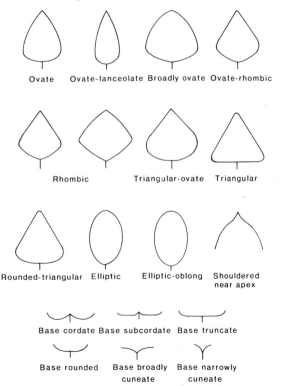

**Figure 3.1** Terms used to describe leaf shape and leaf bases

clearly understood diagrammatic shapes have been provided in Figure 3.1. It is hoped that the drawings will prove to be of assistance to those using the key.

The leaf characters in the key are intended to apply to mature foliage on the lower third of long shoots of medium vigour taken in the second half of the summer. The leaves produced on rapidly growing apical shoots in the upper crown, on epicormic shoots arising from the stem, on coppice shoots and on young nursery stock are often quite abnormal, and this key cannot be used for such material.

Many new cultivars have been released to growers in Europe during the last two decades that are excluded from this key. Some of them may shortly be released to growers in Britain and, when they are, caution will be required in attempting identification of poplars planted since the publication of this Bulletin. Some traditional poplar clones may be encountered in the landscape that are not included in this key.

## Key

1. *Leaves* near the apex of long shoots, together with the shoots themselves and buds, clothed beneath especially when young with white or greyish-white felted hairs;
   *petioles* (leaf-stalks) not conspicuously flattened;
   *buds* ovoid;
   *bark* white or yellowish, ringed with dark, lozenge-shaped scars, furrowed only at the base of adult stems;
   *catkin-scales* persistent, fringed with long hairs;
   *seed-capsules* long and slender, narrowly-conical **2 (white poplars)**
   Some or all characters different from the above **4**

2. *Long-shoot leaves* more or less persistently white-felted beneath, palmately 3–5 lobed like those of a maple (though often less deeply so);
   *short-shoot leaves* not lobed but very coarsely toothed, with little or no felt **3**
   *Leaves* not lobed but coarsely toothed only, their felt greyish and disappearing gradually with age;
   *crown* spreading and very leafy, the finer *branches* more or less pendulous
   **P. × canescens**

3. *Crown* spreading **P. alba**
   *Crown* fastigiate, resembling that of *P. nigra* 'Italica' but broader **P. alba 'Pyramidalis'**

4. *Leaves* all roundish (except on suckers and coppice), coarsely toothed, usually rounded or bluntish at their apex, sometimes clothed with silky hairs when young but quickly shedding them;
   *petioles* conspicuously flattened;
   *buds* ovoid, not hairy, dry or only slightly sticky;
   *bark, catkin-scales* and *seed capsules* resembling those of the preceding species, *P. alba* **P. tremula**
   *Leaves* usually without hairs on the undersurface, not conspicuously hairy, with numerous finer (sometimes obscure) and more or less regular teeth;
   *petioles* flattened or round;
   *buds* elongated, narrowly conical, sticky, usually not hairy and never conspicuously so;
   *bark* usually furrowed for some distance up the stem in adult trees;
   *catkin scales* quickly shed, not fringed with long hairs, though usually lacerated;
   *seed capsules* more or less ovoid **5**

5. *Leaves* green on both sides, though paler beneath, with a clearly defined translucent cartilaginous margin and often obvious toothing;
   *petioles* conspicuously flattened towards their apex;
   *buds* sticky but not very scented **6 (black poplars)**
   *Undersurface of leaves* with a whitish, greyish or greyish-green metallic appearance, sometimes flushed with a rusty coloration, without an obvious translucent margin, often quite obviously toothed;
   *petioles* roundish in cross-section, not conspicuously flattened;
   *buds* sticky and strongly scented
   **17 (balsam poplars)**

6. *Junction of petiole and blade* of all *leaves* without glands;
   *leaf margin* not ciliate;* base of *leaf-blade* usually cuneate in outline;
   *branchlets* round in section, not angular or ribbed **7**
   *Junction of petiole and blade*, at least on some *leaves*, with one or more glands;
   *leaf margin* ciliate;*
   *branchlets* often angular or ribbed, especially when vigorously grown **10**

*The ciliation, when present, consists of short incurved hairs situated actually on the margin and properly visible only under a lens. Care must be taken not to mistake for ciliation the longer hairs which may exist on the leaf surface and project beyond the margin (e.g. in *P. nigra* var. *betulifolia*).

7.  *Crown* spreading;
    *stem* usually with burrs      8
    *Crown* fastigiate;
    *stem* usually fluted      9

8.  *Branchlets* and *leaves* without hairs     **P. nigra**
    *Young branchlets*, and often *leaves*, hairy
                        **P. nigra var. betulifolia**

9.  *Trees* male only;
    *leaves* and *branchlets* without hairs;
    *leaves* usually small and strongly rhombic, often broader than long;
    *crown* strictly fastigiate     **P. nigra 'Italica'**
    *Trees* male only;
    *leaves* and *branchlets* hairy;
    *leaves* usually small and strongly rhombic, often broader than long;
    *crown* strictly fastigiate     **P. nigra 'Plantierensis'**
    *Trees* female;
    *leaves* and *branchlets* without hairs;
    *leaves* larger, usually more triangular and longer-pointed;
    *crown* broader and looser     **P. nigra 'Gigantea'**

10. *Glands* at junction of petiole and blade always present, two or more in number;
    *leaves* usually longer than broad (8–14 cm long), sometimes shouldered near the apex and usually narrowed at the apex to a slender acute point, truncate to shallowly cordate at the base;
    *branchlets* angled;
    *catkin-scales* only irregularly notched, not lacerated     **P. deltoides**
    *Glands*, when present, one or two in number but absent on many of the leaves;
    *leaves* mostly about as long as broad, not shouldered near the apex, smaller;
    *catkin-scales* deeply, irregularly, palmately lacerated     11

11. *Young branchlets* with minute erect hairs (visible only under a good lens);
    *petiole* upper surface with minute erect hairs (leaves from strong shoots);
    *leaves*, appearing very early (see Table 3.1) and of a strong bronze-red when unfolding, rather large (about 10 cm long and broad on strong shoots), their base rounded, truncate or shallowly cordate;
    *crown* narrow and regular, densely leafy;
    *trees* male only     **P. 'Robusta'**
    *Young branchlets* and *petioles* without hairs;
    *leaves* and *crown* resembling those of the preceding cultivar, *P.* 'Robusta'     **P. 'Heidemij'**
    *Young branchlets* quite hairless;
    *leaves* mostly smaller than in *P.* 'Robusta';
    *trees* male or female      12

12. *Base of leaf-blade* mostly broadly cuneate to truncate to subcordate or shallowly cordate     13
    *Base of leaf-blade* mostly cuneate to broadly cuneate      16

13. *Trees* male only      14
    *Trees* female only      15

14. *Leaves* appearing late, one to two weeks ahead of those of the following cultivar *P.* 'Serotina', bronze when unfolding then light to mid-green when mature;
    *leaf-apex* comparatively long, narrow and sharply twisted;
    *twigs* light brown;
    *stem* straight, less lean than in *P.* 'Serotina';
    *crown* usually symmetrical;
    *branches* slightly sinuous, ascending to begin with then curving, downwards or where continuing to grow upwards, branchlets are at wide angle to vertical (upper crown)     **P. 'Gelrica'**
    *Leaves* appearing very late (last poplar to leaf), bronze when unfolding then dark green when mature;
    *leaf apex* broad, short and not twisted;
    *twigs* medium to dark brown;
    *stem* straight but usually leaning;
    *crown* symmetrical when young, becoming asymmetrical with age;
    *branches* ascending to begin with except in lower crown then curving upwards so that young branchlets are nearly vertical     **P. 'Serotina'**
    *Leaves* appearing very late (similar in size and shape to those of the preceding cultivar *P.* 'Serotina') but yellowish-green to yellow when mature     **P. 'Serotina Aurea'**
    *Leaves* appearing very late and similar in size, shape and colour to those of *P.* 'Serotina';
    *stem* straight and usually vertical;
    *crown* symmetrical, strictly fastigiate and a near cylinder in outline;
    *branches* short and at same angle to vertical     **P. 'Serotina de Selys'**

15. *Leaves* broadly cuneate or truncate, never cordate at base;
    *stem* straight, usually leaning;
    *crown* spreading sideways with age and becoming asymmetrical;
    *branches* sinuous and at wide angle to stem     **P. 'Regenerata'**

*Leaves* shallowly to strongly cordate, never cuneate at base;
*stem* straight and vertical or nearly so;
*crown* broad but remaining symmetrical;
*branches* ascending throughout their length
*P.* 'I-78'

16. *Trees* female only;
*leaves* broadly cuneate at base but towards shoot apex base becomes shallowly cordate, often as broad as long (8 cm);
*leaf apex* rather long and twisted;
*stem* sinuous and leaning;
*crown* wide and rounded;
*branches* sinuous, ends pendulous;
*twigs* pendulous upturned at extremities
*P.* 'Marilandica'

*Trees* male only;
*leaves* broadly cuneate at base, none cordate, length usually greater than width;
*leaf apex* broadish and not twisted;
*stem* straight, vertical, often into leading shoot;
*crown* narrow and symmetrical;
*branches* ascending to begin with then curving gently downwards;
*twigs* not pendulous
*P.* 'Eugenei'

17. *Crown* fastigiate, narrow;
*branches* short and strongly ascending at ends, few pendulous twigs
*P.* 'Balsam Spire'

*Crown* not fastigiate, sometimes spreading;
*branches* usually long and with pendulous twigs
18

18. *Young branchlets* normally round     19
*Young branchlets* normally ribbed or angled     21

19. *Young branchlets* and *petioles* not obviously hairy; leaves rather leathery in texture, broadly ovate to ovate-lanceolate, rounded to broadly-cuneate at the base
*P. balsamifera*

*Young branchlets* and *petioles* obviously hairy     20

20. *Leaves* wrinkled, rather leathery in texture, elliptic-oblong or roundish with a subcordate base and an abrupt twisted point;
*petioles* usually 2.5 cm or less in length;
*catkins* (female) long (18–25 cm), conspicuous until September–October;
*trees* male or female
*P. maximowiczii*

*Leaves* smooth, rather leathery in texture, broadly ovate and rounded to markedly cordate at the base and narrowing to a straightish flat point at the apex;
*petioles* longer than 3 cm;
*catkins* long (16 cm) with conspicuous yellowish-green stigmas;
*trees* female only
*P. candicans*

21. *Leaves* dark green above, very grey or white beneath, ovate to narrowly ovate with a rounded, slightly cuneate or slightly cordate base, their margin not translucent, with teeth often obscure;
*buds* very sweet scented;
*catkin-scales* minutely hairy but not fringed with long hairs
*P. trichocarpa*

*Leaves* not very grey/white beneath, lighter green above and thinner than in *P. trichocarpa*, more ovate-rhombic to rounded-triangular in outline, with a very narrow translucent margin and more evident teeth than in *P. trichocarpa*;
*buds* not particularly sweet-scented;
*catkin-scales* without hairs     22

22. *Leaf-base* cordate to truncate;
*leaves* large (7–12 cm, often larger on young trees), pale green beneath, triangular-ovate, with teeth often well marked;
*petiole* and midrib often flushed with red;
*branchlets* hairless or with sparse minute hairs
*P.* × *generosa*

*Leaf-base* rounded to cuneate;
*leaves* mostly smaller (4–10 cm), rather pale greyish green beneath, ovate to ovate-rhombic, finely toothed;
*petiole* with scattered hairs to begin with;
*branchlets* often rather hairy to obviously so
*P.* × *berolinensis*

## Sex and date of leafing

To identify a poplar correctly it is usually necessary to know its sex. This can be discovered by examining catkins at the appropriate time in the spring. But during the winter, catkin buds are far enough advanced for the sex to be determined by dissecting them under a hand lens. Another character providing considerable help in differentiating some cultivars, in particular the black hybrids (*P.* × *euramericana*), is the order in which foliation occurs. Of course the actual dates vary quite widely from year to year and in different parts of the country, but where a number of poplars are growing together under comparable conditions their order of leafing is usually the same. In Table 3.1 poplars are listed in five groups in order of leafing. Within each group species and cultivars are shown in their most likely order of foliation, but this cannot be regarded as constant and in some years a clone may leaf so early or so late as to fall in the group above or below the one it occupies in the Table. It is certainly not possible to identify poplars by relative date of leafing alone, but given a knowledge of the sex of the tree, its habit and its summer foliage and twigs, date of leafing may be the deciding factor.

Several species and hybrids – notably *P. tremula*, *P. alba* and *P.* × *canescens* – show such a wide range of variation in time of leafing that it is not possible to include them in the table. The many forms of *P. nigra* also vary in foliation date but some cultivars of this species have been included. It may be noted that the balsam poplars in general are very early into leaf, those of Asian origin leafing before the North American species and hybrids. Some Asian poplars, for example *P. koreana*, may come into leaf as early as March, clearly an undesirable trait considering the vagaries of the British climate. Hybrids having *P. trichocarpa* as a parent are also very early into leaf. Most poplars come into leaf during April in the southern half of England, however, though in an average season *P.* 'Serotina' may only just be coming into leaf at the beginning of May. When date of leafing is to be used as an aid in identification, the last week in April and the first in May is probably the best period. The earliest poplars are by then fully in leaf and can be distinguished by their complete covering of foliage; slightly later species and cultivars will have the blades of the earlier leaves fully expanded but will still look sparsely clothed. The mid-season and late poplars will still be in different stages from leaves unrolling to buds swelling.

**Table 3.1.** Date of leafing and sex of poplars

| Comparative period | Species/Cultivar | Sex |
|---|---|---|
| Very early | *P. maximowiczii* (and other Asian balsam species) | M and F |
| | *P. candicans* | F |
| | *P.* 'Balsam Spire' | F |
| | *P. balsamifera* | M and F |
| | *P.* × *generosa* | M and F |
| | *P. trichocarpa* | M and F |
| Early | *P.* × *berolinensis* group | M and F |
| | *P.* 'Robusta' | M |
| | *P.* 'Heidemij' | M |
| | *P. nigra* 'Italica' | M |
| Mid-season | *P.* 'I-78' | F |
| | *P.* 'Marilandica' | F |
| | *P.* 'Eugenei' | M |
| Late | *P.* 'Regenerata' | F |
| | *P. nigra* 'Vereecken' | M |
| | *P. nigra* 'Manchester' | M |
| | *P. nigra* 'Gigantea' | F |
| | *P.* 'Gelrica' | M |
| Very late | *P.* 'Serotina' and its varieties | M |

# Chapter 4
# Choice of Site

## Climate

Most poplars are extremely responsive to climate, particularly to summer temperatures. Within reasonable limits, the hotter the summer the faster poplars will grow, so long as soil moisture is available. In addition, most poplars have a long growing season. In Britain, though growth to begin with is rather slow, shoots elongate with increasing rapidity from mid-June onwards, a peak growth rate is reached between mid-July and early September, and resting buds are not usually formed till towards the end of October. The time and duration of maximum growth rate can vary considerably, especially from one poplar to another. But the rate can also be markedly affected by factors other than temperature. For instance, intensity of pruning can significantly influence radial increment, the more severe the pruning the slower the rate of growth (Jobling, 1963/64).

In countries where poplars are much more vigorous than in Britain there is a direct relationship between summer temperatures and increment, at least on optimum sites where adequate supplies of soil moisture are available irrespective of rainfall. Where periods of drought are associated with hot summers, the effects of higher temperatures are partially or even wholly off-set by lack of water. In southern Europe, where poplars increase in size much more rapidly than in Britain, the improved increment is due in part to earlier leafing and a later cessation of growth as well as to the higher summer temperatures.

The relationships between climate and rate of growth in Britain and, in turn, choice of species and cultivar, is best considered by dividing the country into three regions:

1. That part of England lying south of a line from Chester to York but omitting the southern Pennines,
2. The lowland areas of north England and of Scotland and Wales,
3. The Lake District, the Pennines, the Border Hills, the Highlands of Scotland and upland Wales.

In general, the region south of Chester and York (1) is the warmest and driest of the three. In reality only the southern half of England may be regarded as part of the major region in northern Europe, which stretches from southern Sweden to northern France, where black poplar hybrids have been extensively cultivated for nearly three centuries. On the basis of summer temperatures England is probably below the average for this region; on the basis of rainfall and length of season it is probably average or slightly above. In the lowland regions of northern and western Britain (2) summer temperatures and length of season are both below the desirable figures for black poplar hybrids. This applies with even greater force to all the hilly regions of the country (3), where the higher rainfall tends to be reflected in an even greater depression of the summer temperatures.

The southern region (1) can be regarded as perfectly suitable climatically for the cultivation of black poplar hybrids, although rates of growth are always likely to be lower than in warmer climates abroad. The northern and western lowlands (2) will also support euramerican hybrids, but growth is slower and there is a tendency for crowns to be rather sparse and for a certain amount of dieback to occur. Nevertheless, fine healthy black poplar hybrids can be grown as far north as Inverness. In the hilly regions of Britain (3) euramerican hybrids usually grow slowly and unhealthily, or even fail altogether, though occasionally some of the older hybrids such as 'Regenerata' and 'Serotina' are found growing quite well on the most sheltered and fertile sites.

Aspen, *P. tremula*, occurs naturally in all three regions but large, healthy trees are quite rare everywhere. The finest specimens are to be seen in the Highlands of Scotland. Grey poplar, *P.* × *canescens*, grows well everywhere in Britain, though it attains its greatest dimensions on lowland sites in the south. The balsam poplars, as long as they remain free from bacterial canker, also grow well in all parts of the country. They appear to appreciate the high rainfall common to the hilly regions. Until recently the finest specimens of *P. balsamifera* and *P. candicans* were to be found in the shadow of the Grampian Mountains in Scotland, while the largest trees of *P. trichocarpa* thrived in Snowdonia in North Wales.

All the evidence suggests that selected euramerican and balsam poplar cultivars can be successfully cultivated in the Midlands and south of England. But the safe extension of poplar growing to the lowland areas of Scotland, Wales and northern England, and especially to the hilly parts of Britain depends on the availability of balsam poplars. Clones of *P. trichocarpa* or of *P. trichocarpa* hybrids which combine resistance to bacterial canker and climatic adaptability are the only suitable prospects.

## Soil and soil moisture

The ideal site for poplar is a base-rich loamy soil in a sheltered situation, with the water table 1–1.5 m below the surface in the summer. But in practice most poplars can be cultivated successfully on a wide variety of soils ranging from clays to sands. They can also be grown on the alkaline peats of the fens and carrs of East Anglia, provided they are properly drained. Acid sphagnum peats appear on present evidence to be quite unsuitable, however, and the light sand and gravel heath soils of south and east England are often too dry as well as too acid. In contrast, the heavy clay soils in the south Midlands and Thames Valley are quite suitable, though the growth of most species and cultivars tend to be slower and the trees of some may never reach large dimensions.

In general pH value of a soil reflects its suitability for poplar, although other factors such as soil depth and drainage also have to be taken into account. As a rule, a soil with a pH in the range 5.0–6.5 may be expected to support satisfactory growth of all species and cultivars. When the pH is less than about 5.0 there is usually a reduction in the vigour of black poplars and their hybrids, and in soils more acid than this balsam poplars and aspens should be preferred. But when the pH is as low as 4.5 even these relatively acid tolerant trees should be avoided. Where wavy hair-grass (*Deschampsia flexuosa*), mat-grass (*Nardus stricta*), or purple moor-grass (*Molinia caerulea*), are dominant or even prominent on a site, or where the heaths, *Calluna* and *Erica* spp., rhododendrons or bilberry (*Vaccinium myrtillus*) are well established it is likely the soil is too acid for any poplar.

An increase in pH value to 7.0 has little or no effect on most poplars but in alkaline soils (pH>7.0) there may be detectable reductions in vigour. Balsam poplars in particular dislike markedly alkaline soils and some quickly display lime induced chlorosis symptoms. Research in the Netherlands has disclosed that certain balsam poplars sensitive to high pH values show a reduction in foliar nitrogen concentration and iron deficiency symptoms, as well as reduced growth rates.

Poplars, like willows and some alders, are usually associated in Britain with damp soils. As a consequence they are often chosen for wet, badly drained or even waterlogged sites which, in reality, are wholly unsuitable. Poplars are of course hygrophilous trees and are believed to absorb more water through their root systems than other temperate trees. Also, they take up more water relative to dry matter production than other trees. The aspens probably have the lowest moisture demands of any poplar. Yet although their roots certainly require access to a constant supply of water, there are several types of wet site that should not be contemplated at all for planting and others that should be intensively drained before planting can be undertaken. For instance, where there are stagnant conditions – that is where water retreats only as a result of gradual seepage and evaporation – or where the water table never falls below 50–60 cm the use of poplar can be entirely discounted. Whether sites that appear to be less wet than this can be brought into a fit state for cultivation by drainage depends largely on the lie of the land. If there is a reasonable outfall for drains, so that the winter water table can be lowered to at least 60 cm below the surface, poplar can be planted. But if this is not possible however numerous the drains, cultivation will still be out of the question. Flooding in Britain is usually of short duration and does no harm provided the water runs off the site as the level of the river falls.

Land with a heavy growth of reed (*Phragmites communis*), reed-grass (*Phalaris arundinacea*), sedges (*Carex* spp.) or other highly water tolerant plants indicative of marshland or even shallow water is definitely unsuitable. Vigorous growth of marsh marigold (*Caltha palustris*), wild iris or yellow flag (*Iris pseudacorus*) or great water dock (*Rumex hydrolapathum*), usually indicates that the water table is too high. Areas carrying a high proportion of meadow-sweet (*Filipendula ulmaria*) are also open to doubt. Whether such sites can be adequately drained for poplar rests upon local topography rather than the vegetation.

When it comes to choosing drier sites for poplar, it is far more difficult to decide what land is suitable and what is unsuitable. However, leaving out the shallow dry sands and heaths such as occur in parts of East Anglia, Dorset, Hampshire and in Surrey, it is safe to say that poplars can be grown on any soil including *deep* sand that could be used for agricultural crops if cleared and ploughed.

## Situation

Most of the best poplar ground lies on the alluvial soils in river valleys. The actual area of such land in Britain is considerable but owing to the prior demands of agriculture, the amount available for poplar cultivation is comparatively small. A certain amount of the ground is at present too badly drained for agricultural purposes; but if it could be drained sufficiently to allow poplar planting, it would then be good enough for farm crops. However, where it has been colonised to form fen wood or carr-willows, common alder (*Alnus glutinosa*), hairy birch (*Betula pubescens*) and buckthorn (*Rhamnus catharticus*) are often prominent components – it might not be worth draining and clearing for agriculture but it might reasonably be drained and used for poplar cultivation. Apart from existing woodlands there are often odd corners which, because of difficulty of access or for some other reason, are unsuitable for agriculture and could be planted with poplar, bearing in mind that poplar can be grown in lines, in small groups or individually, as well as in plantations.

Throughout much of Europe poplar growing to produce merchantable wood has long been associated with agriculture and, though farming patterns have tended to change in recent years, lines of poplar are still commonplace along field boundaries. In Britain, in contrast, farmers on the whole have not been inclined to adopt systems of poplar cultivation readily accepted by their continental counterparts, even though lines of Lombardy poplar and other common cultivars have been and are still frequently employed as windbreaks, usually with the trees at close spacing, and to demarcate farm and field boundaries. In the 1950s, when approved poplars attracted Forestry Commission planting grants for the first time, a special grant-aid scheme was launched to encourage planting in lines. The response by growers was found to be rather restrained, however, when comparisons were made with the level of poplar planting taking place in blocks, and the scheme was soon dropped.

The extent and severity of root competition on adjacent arable crops will depend largely on the soil type and the size to which the poplars are allowed to grow. On deep loamy, sandy, or alkaline peat soils where moisture and nutrients are amply supplied the effect of poplars roots is likely to be much less severe than where soils are shallower or moisture and nutrients tend to be limiting. The frequency of arable cultivations is an important consideration; provided ploughing and other cultivations are repeated annually the extent and size of poplar roots within cultivation depth can be restricted satisfactorily, however attempts to reintroduce arable cultivations after several years under, say, a grass ley is likely to cause unacceptable damage to large poplar roots. On permanent grassland the effect of line planted poplars will be negligible and the trees may indeed provide beneficial shelter for stock. The shading effect of poplar planted in lines will depend on the spacing between trees and the orientation of the line planting. There will be lesser effect to the south side of wide spaced poplar and greater effect on the north side of close spaced poplar when the lines are not east–west. For north–south orientation of lines the effects are likely to be intermediate between these extremes. In the first few years after planting the effect of shade will be insignificant but as the height of the poplar increases some reduction in crop yield, or delay in ripening may occur.

Poplars can also be planted in lines along the sides of rivers and drains, a position in which they are likely to do well, especially if the water is moving. It is advisable to plant the trees a short distance away from the edge to improve their root anchorage, which otherwise would be very one-sided. But they should not be planted on river banks, in places where the water level of the river is sometimes above the surrounding land. The risk of the trees blowing over when the river is high and the bank is sodden, and thus breaking the bank, is too great.

Sites which are marginal from the drainage point of view can sometimes be partially utilised by planting poplar only along the drains. The trees will have the benefit of some movement of soil water and of a slight raising of the ground surface because of spoil thrown up from the drains. These conditions apply even more forcibly to rivers, since it is often possible to plant a single row of poplars along a river or stream bank, although the land behind may be undrainable. Some caution is advisable in a situation of this kind, however, since if the trees lean over the water, felling may be troublesome and, if a tractor cannot be used on the boggy ground, extraction may well present equally difficult problems. Since by-laws may prohibit the planting of trees close to any water course for which a water authority or council is responsible, advice should be sought from the appropriate body before poplar is planted in such situations. Usually, the provision of working space for drainage equipment is a major consideration.

Except on clay soils poplars can be safely planted along roads. In clay soils poplars are liable to cause shrinkage of the soil due to excessive extraction of water during the summer and, consequently, to damage the road foundations. Planting should therefore be kept some distance away from a road where there is a risk of shrinkage. In any type of soil, poplar roots are capable of travelling long distances and may pass underneath a road. It is not uncommon for certain species and cultivars, notably grey poplar, to sucker on the opposite side of the road from the parent tree.

Obviously such vigorously growing trees as poplars should not be put near buildings, lest their roots cause direct damage to the foundations. A distance of 20 m between tree and building is advisable but, taking into account the height to which most poplars can grow and the risk of wind breakage as they get old, other considerations may dictate a greater distance than this. On shrinkable clay soils the question of shrinkage again arises and is generally thought that poplars, along with elms, oaks and willows, are the most troublesome trees. Though damage to buildings is most likely to occur when the trees are within about 15 m of the foundations, poplars are known to cause damage up to a distance of 30 m (Cutler and Richardson, 1981). Clearly much thought should be given to the siting of poplars in all small gardens, as well as in parks and public open space in towns, where there are buildings close by. In shrinkable clays it may be best not to plant poplars at all where there are doubts about the 'safe' distance between the tree and the building. A 'safe' distance is sometimes taken to be the height of the tree at maturity (Anon., 1985). Some, though not all, poplars may reach a height of 28 m in shrinkable clay.

When a building has to be erected near poplars or other trees which, for one reason or another, will be retained after the structure has been completed, steps can be taken to prevent subsequent damage. Two practices are commonly adopted by the construction industry. First, the depth of the foundations can be increased to allow for movement of soil due to shrinkage. Obviously this is best carried out at the beginning of building, though in some instances foundations can be strengthened later. Secondly, an impermeable root barrier, usually made of concrete, can be constructed to a prescribed length and depth in the soil to prevent roots extending physically in the direction of the foundations. A barrier can of course be built in a lot of cases at a later stage. Sometimes raft foundations are employed where conditions permit. In any event it is extremely important to follow technical guidelines such as those prepared by the National House Building Council.

Most poplars, because they are easily established, have a fast rate of growth and reach large dimensions, are useful for amenity. Those which have a fastigiate habit or possess a light coloured bark or decorative catkins or foliage are especially advantageous in mixed plantings in parkland or public open space since they contrast well with other trees. The narrow crowned cultivars in particular are well suited for planting in position where space is limited – where the crowns of wide-spreading specimens would soon begin to meet and interfere with the crowns of nearby trees. Narrow crowned poplars should still be used with the same discretion shown to other species and cultivars, however, since their roots may become just as extensive.

Poplars are also useful for screening factory development sites, industrial works and mineral extraction operations. Sometimes they may not be the preferred trees from the landscaping or ecological point of view, but because they are usually established without much difficulty and are relatively vigorous from the start, they may still have a valuable role to play while slower though more desirable trees begin to grow well and make a visual impact. They can also be planted for shelter, either in belts with other species or pure, at close spacing, in single row windbreaks. Fastigiate cultivars are well suited for single row windbreaks in nurseries or on farms since their narrow crowns intrude less than those of other trees on to cropped land and they recover well from periodic trimming. Early flushing *P. trichocarpa* clones and hybrids are particularly useful for protecting fruit crops, notably young growth and blossom in spring from cold winds. But because of the risk of roots affecting crop productivity and interfering with soil cultivation, poplars are best planted along field and nursery boundaries rather than between beds or within orchards. There is no evidence of poplars blocking open drains but they should not be planted near tile or mole drains.

In western and central Europe poplar planting in woodland has been the exception rather than the rule. This is due primarily to the long dependence on black poplar and its hybrids which, because they are intolerant of competition both above and below ground, hardly ever grow as well in woodland as in pure, wide spaced crops, groups or rows. Some of the problems were identified in Britain in the 1950s and 60s when sustained efforts were made to replant unmanaged woodland composed largely of scrub and coppice with selected black poplar hybrids. Though perhaps the least difficult woodland environment for these trees, their establishment proved to be rather difficult and growth rates were low for the first few years. Competition for moisture and side shade were considered to be main factors affecting their behaviour. In some instances, to avoid the risk of depriving the poplars of light, rigorous cleaning of coppice had to be undertaken for several years. Where cleaning was discontinued after a while, the poplars, because they cast such a light shade, were slow to suppress the woody understorey and in most cases never did so completely. In parts of Germany, black poplar hybrids have been cultivated successfully in mixed woodland for much of this century, usually on extended rotations. Sometimes the poplars were introduced into established woodland where large gaps had been created in the canopy; in other cases the poplars were included in the mixture at planting. Except on a small scale in Austria, neither system has been adopted elsewhere.

In contrast, the aspens (notably *P. grandidentata* and *P. tremuloides* of North America and *P. tremula* of Europe) and the two major balsam poplar species of North America (*P. balsamifera* and *P. trichocarpa*) are typical forest trees and form dense pure stands or mixtures with conifer and other broadleaved species. In Britain *P. tremula* occurs naturally in woodlands in a wide range of soils and situations often where other poplars cannot be grown successfully. It grows better and is much more common in the Highlands of Scotland than in lowland England but never attains the growth rate or dimensions of other poplars. It is a much less exacting tree than *P. nigra* and its hybrids. Interspecific aspen hybrids bred in Scandinavia and North America have proved in trials in this country to be faster growing and capable of reaching a larger size than the native *P. tremula*. But they have been far too susceptible to bacterial canker to deserve general use. If disease resistant hybrids could be produced, displaying greater vigour and attaining larger dimensions than *P. tremula*, and showing the same tolerance of difficult soils as this species the commercial range of poplars could be greatly extended in this country.

The valuable role of *P. trichocarpa* in British woodlands is much clearer due to the increasing availability in recent years of vigorous, canker resistant clones of this species as well as its hybrids. They have undoubtedly led to prospects of successful poplar cultivation on sites which, hitherto, have been considered unsuitable for poplar. They grow in more acid soils than the black poplars, they stand more competition than these exacting trees and they can be safely planted in the wetter and cooler parts of the country. It is encouraging to recall perhaps that for several decades after its introduction at the turn of the century, *P. trichocarpa* grew impressively in woodlands in Scotland and North Wales until seriously attacked by bacterial canker. It is also notable that trials to evaluate the place of *P. trichocarpa* in forests are advanced and widespread in Germany.

# Chapter 5
# Plant Production

Most poplars raised commercially, whether for wood production or for amenity, are reproduced vegetatively in open nursery beds from hardwood cuttings taken from ripened shoots. The few species and hybrids which are difficult to propagate from hardwood cuttings are raised from root cuttings, from root suckers or from leafy softwood or semi-ripe cuttings, taken from current shoot growth. Reproduction from seed is undertaken mainly by tree breeders and by ecologists and others concerned with the conservation of native tree populations. Plant production from seed is hardly ever resorted to in nurseries in Britain.

These methods of propagation are described below. Where appropriate, information is included on varietal control, sources of reproductive material and size specifications for planting stock. Grafting and budding techniques are normally the preserve of tree breeders and are not discussed. There is in any case a strong argument for cultivating only trees on their own roots.

## *Hardwood cuttings*

All poplars in the *Aigeiros* section, excepting some forms of *P. deltoides*, and all *Tacamahaca* poplars are readily propagated from hardwood cuttings. In the section *Leuce*, the white poplars are raised from hardwood cuttings, but not the aspens or hybrids derived from them. Poplars in the section *Leucoides* are difficult to propagate from hardwood cuttings and other methods of reproduction are usually adopted.

Cuttings are prepared from well-ripened, fast grown one-year-old shoots. Where a large annual supply of material is required, the shoots are produced on stock plants or stools established at a wide spacing of at least $1 \times 1$ m in a fertile nursery bed and cut back to the ground each winter. Otherwise cuttings are taken from 1-year-old rooted cuttings cut back in the winter, to just above soil level. Cuttings from vigorous pollards, coppice and epicormic growths are acceptable; but cuttings from the crowns of mature trees should be avoided as, in general, should cuttings taken from two-year-old wood.

There appears to be a relationship between size of cutting and root and shoot growth in cutting beds. In nurseries in Britain cuttings are usually 20 to 25 cm long, with a diameter of 10 to 20 mm at their midpoint. Smaller cuttings than this will root but the plants tend to be weaker. Cuttings with a diameter less than 10 mm may even fail to root. The top cut end of the cutting should be about 1 cm above a leaf bud, the bottom end close to or just below a bud. Cuttings which have no obvious leaf buds will often throw out shoots from adventitious buds, but as there is a delay before growth starts, the resulting plants may be small. Thick cuttings from the basal parts of shoots can be used, so long as they can be inserted easily into the soil. The top third of shoots is usually too thin to use.

If the nursery soil has been properly cultivated, it should be possible to push cuttings in by hand without damaging the buds or bark. The cuttings are inserted vertically until the top cut end is close to the soil surface. When the soil has settled the tops of the cuttings can then be seen and the rows easily picked out. Though cuttings are prepared throughout the dormant season, nothing is gained by inserting them early in the late autumn–early winter period. In fact, serious rotting of cuttings can occur overwinter in wet soils, resulting in their failure. Insertion is best carried out in late winter to allow some root initiation and development to take place before the first buds open. In practice, cuttings of early leafing poplars in the section *Tacamahaca* should be in the ground by the end of February, and poplars in the *Leuce* and *Aigeiros* sections by the end of March.

Cuttings may be stored in air-tight plastic bags or film in a cold chamber, at 3–4°C, until required. Alternatively they can be kept in a good condition by placing them upright in moist, coarse sand. The cuttings should be completely covered and the sand prevented from becoming waterlogged. Whatever method of storage is employed, the aim is to limit water loss from the cuttings. Dry cuttings do not root.

In fertile, sheltered nurseries most poplars grow vigorously from hardwood cuttings. To encourage the development of tall, sturdy and well-shaped plants, cutting spacings are usually wide. Spacings of about 50 × 50 cm are commonly used though cuttings are sometimes set as close as 30 cm in the rows with the rows 80–100 cm apart. Many new poplar selections released recently by research stations to horticulture and to timber growers have particularly fast rates of growth in nursery beds. The most vigorous are clones of *P. trichocarpa* and crosses derived from *P. trichocarpa* and *P. deltoides*; these require a cutting spacing of 50 × 100 cm. Local site conditions and the type of equipment in use in the nursery may influence choice of spacing as much as the growth habit of the poplars being propagated.

Two or more shoots may arise from a cutting. When this happens the straightest and strongest is selected as the main stem and the others removed. This can be done when the shoots are about 25 cm long. Most plants of the commonly raised poplars can attain a height in excess of 120–150 cm in the first season. Sturdy, well rooted stock taller than 150 cm are usually considered suitable for planting out.

One-year rooted cuttings appreciably shorter than 120 cm are generally retained in the nursery for a further season. Sometimes, the plants are cut back during the winter before the second season starts, leaving two or three buds above the root collar, and the severed stems used as a source of cuttings. The resulting plants, which after two seasons have a 1-year-old top and a 2-year-old root, generally have a straighter and a sturdier stem than in the first season, and a better developed root system relative to the size of the stem and branches.

A system of plant production developed in Italy, and sometimes practised in northern Europe, entails cutting back and transplanting the rooted cuttings at the end of the first season. The cuttings are inserted as close together as 10–15 cm in the rows with the rows up to 100 cm apart; the one-year rooted cuttings, often called barbatelles, are lined-out at spacings not less than 50 cm in the rows or 100 cm between rows. Traditionally, barbatelles are cut-back before being transplanted. However, it is usually more practicable to move rooted cuttings from one nursery bed to another with the stems intact, and cutting-back is undertaken more often than not after the plants have been lined-out. Since transplanting is comparatively expensive, the barbatelle system of plant production has not found favour in Britain. The more economic use of nursery space and an opportunity to grade and cull plants at the end of the first year, the main advantages of the system, hardly offset the high costs of handling and lining-out the one-year rooted cuttings.

Poplars cannot be identified without selected foliage being available. As a consequence, if cuttings, sets or rooted plants of different clones are accidentally mixed between leaf-fall and mid-summer, their separation is likely to prove impossible. The only exception to this rule is when two clones from different sections are mixed. Then, their separation should prove reasonably straightforward. None the less, it will not be possible to identify them without their names being known previously. It follows that all types of reproductive material, as well as rooted stocks earmarked for planting-out, should be carefully handled at all times to maintain the separate identity of different clones. A legible and generally acceptable system of labelling is essential, while clear plans of all stocks in nursery beds should be kept up to date and retained for office reference.

Stool beds which have been planted and managed for the annual supply of hardwood cuttings or the periodic supply of sets should also be well labelled and accurate ground plans maintained in offices. Mixing of material of different clones usually occurs when stools are being cut, or when cuttings or sets are being prepared from stool shoots. At these stages careful labelling of shoots and cuttings is essential. The risk of mixing can be minimised by completing work on one clone before work is started on another.

Under European Community regulations poplar cuttings and plants used for forestry purposes should ordinarily be raised from officially approved reproductive sources recorded in the National Register of Basic Material. Since 1977, more than 50 poplar stool beds in different parts of Britain have been entered in the National Register as approved sources of material. Only beds of cultivars which satisfy criteria laid down in European Economic Community Directives have been accepted. These are cultivars which have proved in practice, as well as in comparative scientific tests, to be capable of vigorous and healthy growth, and of producing merchantable wood free from serious defects. Subject to certain conditions and requirements these cultivars attract Forestry Commission grant-aid when planting stocks are raised from registered stools.

The requirements for the approval of basic material and the conditions controlling the marketing of poplar stocks and of seed and plants of 14 forest tree species within the European Economic Community are explained in Statutory Instrument 1977 No. 891: The Forest Reproductive Material Regulations 1977. A list of names and addresses of owners of registered stool beds can be obtained from the Silviculture Branch, Forestry Commission Research Station, Alice Holt Lodge.

Size standards for planting stock raised from hardwood cuttings are also included in Statutory Instruments 1977 No. 891. The standards have been adopted by member states of the European Economic Community to classify poplars in the Section *Aigeiros*. It is possible that the standards (Table 5.1) might also be applicable to most species and cultivars in the Section *Tacamahaca*, but probably not to white poplars and aspens, or their hybrids, in the Section *Leuce*.

Two sets of standards are in use in the Community. One is for the Mediterranean region, the other for regions other than the Mediterranean region. The standards adopted for the Mediterranean region are for stocks very much larger than those ordinarily produced in nurseries in northern and western Europe and are therefore not shown here.

**Table 5.1.** European Economic Community size standards for plants of the section *Aigeiros* (black poplars)*

| Age | Point of measurement (m) | Diameter (mm) | Height (m) |
|---|---|---|---|
| 0 + 1 | 0.5 | 6 to 8 | 1.00 to 1.50 |
|  |  | 8 to 10 | 1.00 to 1.75 |
|  |  | 10 to 12 | 1.00 to 2.00 |
|  |  | 12 to 15 | 1.00 to 2.25 |
|  |  | 15 to 20 | 1.00 to 2.50 |
|  |  | 20 + | 1.00 — |
| more than 1 year | 1.0 | 8 to 10 | 1.75 to 2.50 |
|  |  | 10 to 15 | 1.75 to 3.00 |
|  |  | 15 to 20 | 1.75 to 3.50 |
|  |  | 20 to 25 | 2.25 to 4.00 |
|  |  | 25 to 30 | 2.25 to 4.75 |
|  |  | 30 to 40 | 2.75 to 5.75 |
|  |  | 40 to 50 | 2.75 to 6.75 |
|  |  | 50 + | 4.00 — |

*Taken from Statutory Instruments 1977 No. 891: The Forest Reproductive Material Regulations 1977.
Also published in British Standards Institution Specification for Poplars and Tree Willows: BS 3936, Part 5.

## Root cuttings

Poplar plant production from root cuttings is rarely if ever practised these days. However, the technique may be used to reproduce selected trees of *P. tremula* and *P.* × *canescens* found in the field, in order to introduce them into clonal collections, and it can be employed to raise small numbers of stocks of these poplars for field planting when misting equipment is not available in greenhouses.

Root cuttings are usually 5–10 cm long. Their diameter is dependent upon the age of stock plant and the part of the root system from which the root segments can be conveniently taken. When stock plants are especially managed in nurseries to supply root segments, cuttings prepared sequentially may vary in diameter from about 5 to 40 mm. The cuttings are inserted vertically, or at a slight angle to the vertical, in a well drained substrate of coarse sand and peat, so that their proximal ends are flush with the surface. The best results may be obtained in a cool greenhouse when the cuttings are inserted in pots or deep seed trays. Then, light watering is needed to prevent drying out. The cuttings may also be sealed at each end before insertion, with a thin coating of wax or other inert protectant, to reduce the risk of desiccation.

The most favourable season for propagating clones in the section *Leuce* appears to be the period October to December. However, success rate is influenced more by clonal characteristics than by season of cutting insertion or substrate and air temperatures. Cuttings that have sprouted and developed roots are bedded out or potted up until large enough for planting out.

## Root suckers

Propagation of poplars from root suckers is unlikely to be attractive to nurserymen faced with the annual production of uniform, vigorous and well rooted stocks. The system is labour intensive and consistently satisfactory results are difficult to achieve because of the unreliable survival and vigour of suckers. The maintenance of a steady supply of good quality suckers suitable for lining out can also be a serious drawback. The technique is only of interest, therefore, when small numbers of plants have to be raised, of poplars such as *P. tremula* and *P. × canescens*. which cannot readily be propagated from hardwood cuttings. Sometimes the method is useful for introducing new selections from the field into clonal collections.

Suckers arise from lateral roots close to the soil surface and spread outwards from the tree as the root system extends. Most poplars have wide-spreading root systems and some suckers may be more than 30 m away from the main stem. As the shoot of a young sucker increases in size, root initiation and development occurs directly below it on the underside of the lateral root. With the regeneration of its own root system the sucker is capable of independent growth and, if carefully severed from the parent tree, it can be transferred successfully to another site.

Usually the size and vigour of a sucker shoot provides little indication of the quality of its root system and, before its severance from the parent tree, the likely behaviour of a sucker cannot easily be predicted. Often, an apparently vigorous sucker is found to have a poorly developed root and it dies soon after planting out. Though suckers are sometimes transferred directly to the field, their chances of survival are much improved when lined out in a well manured and cultivated nursery for a year or two.

The vigour and stem shape of suckers can be improved by cutting back the original shoot to about 5 cm above the soil surface after lining out. Minimising the delay between severance and lining out, and plentiful watering in dry spells during their early weeks in the nursery may improve the survival rate.

Most poplars in the section *Leuce* sucker readily. Severance of roots to permit sucker removal together with other forms of root damage, due to soil cultivation for example, inevitably encourages fresh sucker development. Simple techniques thus ensure the annual supply of small numbers of plants.

## Softwood cuttings

All poplar species and cultivars raised commercially in nurseries in Britain can be readily propagated from softwood (or leafy or summerwood) cuttings in greenhouses equipped with automatic overhead mist irrigation. Semi-ripe cuttings are only slightly less easy to root than softwood cuttings. Poplars in the sections *Aigeiros* and *Tacamahaca* are especially simple to propagate and a 95–100 per cent success rate can usually be achieved throughout the growing season.

As cuttings show root initiation and development after only about two weeks, a monthly turnover of cutting batches can be obtained and some hundreds of plants raised from one stock plant in a single season. With one exception, though, black and balsam poplars and their hybrids are readily reproduced from hardwood cuttings in open nursery beds and, for various reasons, this method is practised by nurserymen. Nevertheless, the softwood cutting technique is a valuable means of rapidly increasing stocks of selected clones.

The best way of propagating *P. tremula, P.* × *canescens* and other poplars in the section *Leuce* is from softwood and semi-ripe cuttings. Cuttings can be taken and rooted throughout the growing season. The fastest and highest rooting rates and the largest plants at the end of the summer are achieved with batches of cuttings inserted in the first half of the season. In the latter part of the season, some falling-off in rooting rates is likely to be experienced and plants are smaller and comparatively poorly rooted at the end of the year.

Cuttings should be sturdy and 10–15 cm long, cut just below a node, taken from vigorous, current growth. A greater success rate may be achieved with apical cuttings than with subapical cuttings, especially in June and July, but care is needed to prevent wilting and death of very soft apical cuttings at the beginning of the season. Lower leaves should be removed from the cuttings before they are inserted in a freely-drained substrate of peat–coarse sand or peat–horticultural vermiculite. The base of the cuttings should be some 2 cm above the floor of the cutting tray to minimise rotting. Treatment with a root promoting growth substance may not significantly affect the number of cuttings rooted but the speed of rooting may be improved. Proprietary powders based on indole-3-butyric acid (IBA) or 2-naphthaleneacetic acid (NAA) dust in talc, at concentrations not greater than 3000 p.p.m., are satisfactory. Treatment with a fungicide to lessen the risk of rotting may be equally or even more beneficial.

Cuttings should be taken for preference from young, vigorous nursery stock plants cut back annually during the dormant season. The plants should be close-to-hand to minimise handling and to lessen delay between collection and insertion. If need-be, cuttings can be stored safely for some days in a cold chamber at +3°C if lightly packed in plastic bags. Cuttings of *P. tremula* and *P.* × *canescens* taken from sucker shoots root less well than cuttings from stock plants in nurseries, but they give better results than material from the crowns of mature trees.

If the temperature of the rooting medium can be raised artificially to and be maintained at 21–24°C, a high proportion of cuttings should have rooted after 3–4 weeks, when a short period of weaning can begin. If substrate temperatures are not kept at this level, root initiation and development may be delayed a week or so. In any event, shading and ventilation are required to prevent rapid rises in air temperature in the greenhouse, and misting should be copious enough to inhibit wilting and failure of cuttings, especially during the critical period immediately after insertion when recovery from handling may be in the balance.

Cuttings can be inserted in trays as close together as 3–5 cm. After weaning, early batches of rooted cuttings should be potted-up as quickly as possible using a well fertilised, peat based compost, and kept in an unheated plastic tunnel until the following season. Because small and comparatively poorly rooted plants from cutting batches inserted towards the end of season overwinter uncertainly after potting-up, such cuttings are best left in their trays, in plastic tunnels, until the following season. Success rates of about 90 per cent can be obtained with *P.* × *canescens* and of 70–80 per cent with *P. tremula*.

## Seed

It is very rare indeed to find poplar stocks in nurseries in Britain which have been raised from seed. This is not the case in some other countries in Europe, where seed collections of *P. tremula* are made annually especially to satisfy horticultural requirements. The seed is gathered from seed orchards as well as natural stands. There is also a long history of both artificial and natural poplar breeding in European research stations, leading to large and continuing programmes of plant production from seed. A first generation hybrid between *P. tremula* and *P. tremuloides*, employing pollen of the latter species imported from North America, has also been raised commercially from seed in very large quantities for more than 30 years. In

Britain there is no poplar breeding and what little seed is collected from time to time, notably of *P. nigra*, is required for taxonomic or conservation purposes. In practice, seed of *P. tremula* is hardly ever seen in Britain.

Seed of *P. tremula* are not difficult to collect or to store, contrary to popular belief. Catkins are collected when the familiar white down or pappus becomes visible at their base. The appearance of the pappus is an indication that ripening of seed has started; ripening begins at the base of the catkin and progresses upwards. After collection, the catkins are kept under cover, in warm greenhouses for preference, until all the fruit are open and the seed can be collected. The catkins may be hung from lines in the greenhouse or the twigs can be placed in containers fed with water at not more than 10°C. High temperatures, shading and low relative humidity help the seed to ripen. Alternatively, the catkins can be spread out on the floor so long as strong draughts can be prevented.

The seed are easily extracted from the fruit using vacuum suction; then, by passing the air through a number of containers the seed and the pappus are separated. The simplest and cheapest method is to place the uncleaned seed in a paper bag and to blow into it. The pappus tends to rise and the seed fall to the bottom of the bag. Unless dried for storage seeds should be sown as soon as they become available. If they are being put into storage they must be dried until their moisture content is reduced to about 5 per cent. Drying should be slow to begin with, for 2 days at room temperature, then rapidly speeded up by using a current of dry air at a temperature of 20–30°C (FAO, 1979). Storage for at least 3 years without loss of viability can be achieved by keeping the dried seed in vacuum packed containers at low temperature.

Sowing in frames or in seed trays in plastic tunnels is preferable to sowing in open nursery beds. After sowing, the seed should be lightly pressed into the soil, but not covered, and watered until germination is established. Shading until the first real leaves appear is considered to be beneficial. Germination begins some 18 hours after sowing but no elongation of the radicle takes place until after the sixth day. In the first week or so heavy rain or careless watering should be guarded against to ensure that the radicle becomes firmly established.

At the end of the first year the seedlings, which may attain a height of 20 cm, are lined out in open nursery beds and cut back to just above the root collar. Cutting back encourages the production of straight stems. In fertile beds, many if not all of the transplants may be tall enough for planting out by the end of the second season. If plants have to be retained in the nursery for a third year they should be re-lined at a spacing of 60 × 90 cm and cut back again. For further information on these techniques and for supporting pictorial evidence a review by Gray (1949) should be consulted.

## *The nursery*

The soil for a poplar nursery should be moist, fertile and easily cultivated. Extreme soil types should be avoided; heavy clays because soil working and plant lining out and lifting are more difficult, light sands because they are likely to be too dry during the growing season; indeed, the soil needs to be moist at rooting depth throughout the growing season. The soil pH should be between 5.0 and 6.5. The nursery should be well sheltered and free of the risk of both early and late frosts.

Cutting and transplant beds should be fallowed every 2 or 3 years. Fertility should be maintained by regular manuring, especially organic manures at up to 50 t ha$^{-1}$ every other year. Phosphorus and potassium fertilisers are usually required annually or every other year, at the rate of 50–100 kg of $P_2O_5$ and 100–200 kg of $K_2O$ per ha. Nitrogen fertilisers, in contrast, should be applied sparingly to avoid excessive growth towards the end of the summer and slow lignification of young shoots. Cultivation should be deep enough to enable cuttings to be inserted to their full depth.

# Chapter 6
# Planting and Establishment

## Planting

Rooted stocks are always pit planted. The optimum depth of the planting hole will depend upon the height of planting stock and local site factors but the two essential factors to be considered are, firstly, the need to plant at sufficient depth for the developing root system to be always in moist soil even under conditions of summer drought and secondly, for stock to be planted at a sufficient depth to ensure stability during the first growing season. The buried portion of the stem produces new roots while the nursery roots remaining after lifting may also regenerate. Accordingly nursery stock up to 2 m in height should be planted to a depth of between 45 cm and 80 cm and plants in excess of 2 m in height may benefit from planting to a depth of 100 cm or more. In Europe, two-year-old stocks can easily be 4–6 m tall. Then, planting holes as deep as 2–3 m are encouraged. In low rainfall areas, especially where soils are highly porous, such deep planting is commonplace and hydraulic buckets are used to make the holes.

Shallow planting is only acceptable on ground that remains wet for much of the year. The roots are then placed in relatively shallow, though wider than normal, holes and covered with a mound of soil built around the base of the stem. This type of mound planting is not a substitute for proper drainage, however, as poplars will never grow satisfactorily on a really wet site. Regardless of size of tree and planting hole, initiation and extension of roots in usually rapid and newly planted poplars rarely need to be staked.

Soil augers are widely used to make planting holes, most mounted on agricultural tractors. Augers with diameters up to 70 cm have been developed especially for poplar planting. By using a mechanical tool of this sort, rates of planting can be increased six to eight times. Problems only arise on heavy, wet clay soil, when the sides of the holes may become polished and inhibit lateral root growth. This problem may be overcome by preparing the holes well in advance of planting, to encourage weathering and crumbling of the sides by winter frosts. In very wet seasons however, prepared holes may fill with water for long periods in winter and early spring. On some woodland sites inconvenience may be experienced because of existing root systems. However, though the efficiency of the equipment may be lessened, its use is still preferable to preparing pits by hand.

Explosives are sometimes employed to make planting holes. Their use appears to be favoured in France, where good results have been obtained in soils with a shallow pan close to the surface (FAO, 1979). Experiments carried out by the Forestry Commission in the late 1950s showed clearly that poplars are more easily established in holes made by explosive than in hand dug pits (Jobling, 1961). A close relationship was found between rate of growth, size of hole and, in turn, size of explosive charge. The greater vigour of trees in holes prepared by explosive was thought to be due, simply, to the larger size of the holes and to the disturbance of soil at their base and sides. However the use of explosives is not seen as a practical alternative to other methods of planting poplars.

The use of unrooted sets for poplar planting in Great Britain has not been widespread and the practice has hitherto given variable results. The traditional method has been to dig a suitably deep pit and to insert the base of the set into a crowbar hole at the bottom of the pit. Sets pushed into uncultivated soil are usually not planted to sufficient depth or are damaged in the attempt. They are unlikely to be well firmed and for some or all of these reasons may root sparingly or not at all and show poor survival and early growth.

An alternative method which has given good results on deep loam soils is to insert the set into a hole bored into the soil with a cylindrical steel bar just larger than the base of the set and to fill the resulting gaps between the set and the sides of the hole with dry sand poured from a suitable container (Beaton, 1988). The hole should be bored to at least 45 cm for sets of approximately 1.5 m length and may need to reach 50–60 cm or more in depth for larger sets. It is essential for new root hairs on the buried portion of the set to develop in completely stable conditions and any movement of the set below ground resulting from failure to fill gaps around the set thoroughly with dry sand will lead to delay in establishment or death of the plant. It is also essential for the developing root system always to be in contact with moist soil even under conditions of summer drought so that depth of planting can be a critical factor in ensuring survival during the first growing season. Provided the sets are planted to sufficient depth and are thoroughly stabilised in the planting hole this method gives reliable results. With a large planting programme the savings in cost and time compared with the use of rooted stock can be very significant.

In mild localities, planting can be carried out throughout the dormant season, avoiding periods of frost. It should usually be completed by the end of the winter before root and shoot activity start. However, in most seasons, planting of the later flushing *P. × euramericana* cultivars 'Eugenei', 'Gelrica' and 'Serotina' may be safely delayed until the beginning of April.

## Weed control

Poplars are highly intolerant of weed competition, and survival and growth rates can be seriously depressed unless steps are taken to reduce or eliminate weeds around the newly planted trees during the first few seasons. On woodland sites shrub and tree growth interfering with or shading the crowns of the poplars must be cut out. Rigorous cleaning is particularly important where *P. × euramericana* cultivars are being grown. Also, reduction in numbers of woody weeds may be needed to lessen competition for moisture and nutrients.

Competing ground flora, especially grasses, must be controlled in the early years. Several herbicides commonly used in forestry to control woodland grasses and herbaceous broadleaved weeds effectively control most ground flora encountered on poplar sites. A pre-planting application and at least two applications after planting are usually required because of the vigour and invasive nature of weeds on poplar sites. Suitable herbicides include glyphosate, paraquat and propyzamide but refer to Forestry Commission Field Book 8 for details. In each case the manufacturer's instructions should be carefully followed to ensure safe and effective application of the chemical. The treated area should be at least 1 m across.

Mulches are also employed to control weed growth after planting. In experiments started by the Forestry Commission in the 1950s, to evaluate different methods of increasing the survival and growth rates of newly planted poplar, mulching consistently improved tree performance more than any other treatment (Jobling, 1960). In most of the experiments – eight were planted on different sites in different years specifically to compare weed control techniques – mulching led to highly significant increases in height increment. In each case improved growth rates continued even after mulching ceased. The most beneficial mulches were composed of locally cut vegetation piled around the tree as soon as possible after planting, and then repaired as required to maintain weed-free conditions for three to four seasons. Imported straw and bracken were found to be reasonably good substitutes. Mulches of sawdust and fresh bark peelings were also tested and shown to control weeds satisfactorily, but they did not promote the same vigour as mulches of cut vegetation. In all the experiments the mulches were at least 1 m wide and 15 cm deep. These are quite acceptable specifications for present day practice. The effects of maintaining mulches 2 m wide and 30 cm deep for five seasons were also observed experimentally. However, mulches of this size were found to be so time consuming and physically difficult to construct and repair that, regardless of any improvement in growth rate achieved, their use in poplar growing was not considered practicable or economic.

Some opaque, durable materials laid flat on the ground and anchored securely around the base of the tree adequately control weeds during the first few seasons. Black polythene, sisal-craft paper, used fertiliser bags and old roofing felt were found in early poplar experiments to be effective for varying periods. Black polythene is now recommended for controlling weeds in newly planted orchards and is also found effective when planting broadleaved trees (Davies, 1987a and b).

Throughout Europe, excepting Britain, soil cultivation before and after planting is practised to the exclusion of all other methods of weed control. Because of the wide spacings ordinarily adopted between trees, agricultural and horticultural cultivators including tractor-drawn equipment can be used in the plantations. Often, poplar sites are ploughed before planting. After planting, cultivation is undertaken at least once a year with disc harrows or rotary cultivators. Inevitably, cultivation damages surface roots; but since wounding encourages root regeneration and growth of new roots the overriding effects are beneficial rather than detrimental. Sites are normally cultivated for about five seasons, though prescriptions vary widely and soil working may continue until canopy closure. Sometimes complete cultivation is practised, in other cases cultivation may be limited to a strip of ground running either side of the rows of trees. The amount of soil working to be undertaken can be influenced by the need to improve access into the plantation to control pests and to carry out high pruning and other cultural work. Cultivation is also associated with inter-row cropping. In western Europe, where spacings now tend to be at least 7 × 7 m, a wide range of commercial crops is raised up to about the seventh year. This aspect of poplar cultivation is discussed in Chapter 11.

Soil cultivation has been practised for so long in Europe and has become such an accepted part of European poplar culture that it is often taken for granted. As a consequence it is difficult to locate reliable evidence in recent technical literature likely to be helpful to growers in Britain. Cost–benefit appraisals of all the different cultivation systems found in the major poplar growing regions, where sites and climate are usually much more favourable to poplar culture than here, might in any case be misleading. Some data from a 9-year-old trial at Longué, 10 miles north of Saumur in the Loire Valley are, therefore, very welcome. These show that 9 years after planting the breast height diameter of poplars in soil cultivated continually since planting was nearly 2½ times greater than that of trees in soil which had never been worked (FAO 1979). The respective diameters were 28.8 cm (worked soil) and 11.7 cm (unworked soil). Moreover trees which had grown slowly for five seasons in unworked soil quickly showed a marked improvement in vigour when cultivation was started in the fifth year and continued for the next four seasons. Between year five and year nine, when their diameter was 19.3 cm, the trees displayed a similar rate of radial increment as trees always grown in cultivated soil.

## *Fertilising*

Forestry Commission experiments planted in the 1950s to compare different methods of establishing poplars showed that growth rates are improved in the first few seasons by applying fertilisers to the newly planted trees. Nitrogen, as ammonium sulphate applied as a top dressing in mid-June in the first season, produced a marked increase in height growth mainly in the first 2 years. Phosphorus, as calcium superphosphate, and potassium, as muriate of potash, both applied at the sides of the pit at time of planting, produced similar or slightly poorer height increases detectable over a 4-year period. The largest and longest lasting responses were found on a moderately acid, somewhat nutrient deficient soil, the smallest on a base-rich, typical poplar soil. Only slight increases in vigour were obtained by increasing the application rate (nitrogen from 113 g to 227 g, phosphorus from 227 g to 454 g and potassium from 113 g to 227 g) and there were no significant interactions between fertilisers.

Where the effects of fertiliser application were compared with growth responses to weed control measures, mulches were found to improve height increment more than any other treatments. The greatest differences in height growth due to treatment were observed on fertile soils supporting dense and vigorous ground flora including grasses, where the use of a fertiliser had the least and mulching the greatest beneficial effect on tree performance. It was concluded that fertilisers are of most value on marginal poplar sites and that, in order to obtain an early and rapid height increase on such sites, a quick acting nitrogenous fertiliser should be applied to trees in mid-summer. There appeared to be no merit in applying fertilisers in combination.

# Chapter 7
# Spacing, Thinning and Pruning

## Spacing

### Spacing and growth
Poplars show the same general trends of response to spacing and thinning as most other tree species, but they are less tolerant than most of between-tree competition. At wide spacing, diameter increment of individual trees is rapid, but yield per hectare is low. At close spacings the converse is found with smaller individual tree diameter but greater yields per hectare. These general conclusions, which are true of all forest species, are illustrated in Table 7.1 which reports data from two poplar spacing trials.

A further effect of variation in stand density is to influence the age of culmination of maximum mean annual increment (MAI). At close spacings and high stand densities MAI will culminate sooner than at wide spacings and lower stand densities.

The cultivation of poplars has probably used these general spacing and growth relationships more widely than any other forest species: poplars are grown at very wide and very close spacings to achieve specific objectives of management.

### Traditional poplar cultivation
In Britain the great bulk of poplars have been established at wide spacings, typically ranging from 5 to 7.5 m.

The wide spacing and cultivation ensured rapid diameter growth until canopy closure soon after mid-rotation age. Thereafter diameter increment decreased markedly near the base of the tree, particularly with *P. × euramericana* cultivars, to a lesser extent with *P. trichocarpa* and hybrids of this species. Even at spacing of 10 m mutual competition may occur towards the end of a 25-year rotation. Such exceptionally wide spacing leads to an unacceptable loss in volume per hectare except on a much longer rotation than 25 years.

Because almost all poplar cultivation has employed these wide spacings, compared with conventional forest spacings of around 2 m, the yield models published show yield classes of between 6 and 12, i.e. up to $12 \text{ m}^2 \text{ ha}^{-1} \text{ y}^{-1}$ maximum mean annual volume increment (MAI). Quite clearly this is not the maximum productivity that poplars can achieve but reflects the wide spacing used. As Table 7.1 indicates, at close spacings yield classes in excess of 20 are quite practicable.

### Use of close spacing
In recent years considerable interest has been shown in cultivation of poplars as a source of biomass and wood fibre. It has long been known that at conventional forest spacings high yields of poplar are possible and Stern (1971, 1972) suggested harnessing this productivity as a source of hardwood pulp. More recently even closer spacings down to just 25 cm have been examined with a view to raising large quantities of biomass on very short rotations of less than 5 years. Experiments laid down in the early 1980s showed quite clearly that with coppice regrowth from young stools at spacings around 1 m, yields in excess of 10 t dry matter per ha per year (theoretical yield class 25 plus) are feasible. Table 7.2 shows data from one of these trials.

### Recommendation
The strong apical dominance and generally good stem form of approved clones of poplar mean that

**Table 7.1.** Yield data from two poplar spacing trials. Plots of comparable top height grouped together to illustrate effects of different spacing (number of trees per hectare – left hand column)

| | | Number of trees | Top height (m) | d.b.h. (dom) (cm) | Mean height (m) | Mean d.b.h. (cm) | Basal area ($m^2\ ha^{-1}$) | Form height | Volume ($m^3\ ha^{-1}$) | MAI (volume) ($m^3\ ha^{-1}\ y^{-1}$) |
|---|---|---|---|---|---|---|---|---|---|---|
| Cobden's copse, Hants. | 19y (a) | 1328 | 21.5 | 26.0 | 21.3 | 21.6 | 48.5 | 9.0 | 437 | 23.0 |
| | | 749 | 21.2 | 32.4 | 20.1 | 27.4 | 44.1 | 7.9 | 348 | 18.3 |
| | | 719 | 21.5 | 32.7 | 21.9 | 28.0 | 44.2 | 8.5 | 374 | 19.8 |
| | | 479 | 21.4 | 36.4 | 20.6 | 32.3 | 39.1 | 8.2 | 321 | 16.9 |
| | (b) | 1301 | 19.3 | 25.5 | 18.6 | 21.5 | 47.1 | 7.4 | 350 | 18.4 |
| | | 1250 | 19.5 | 25.8 | 19.2 | 21.7 | 46.0 | 8.3 | 383 | 20.2 |
| | | 744 | 19.3 | 29.8 | 19.3 | 26.2 | 40.2 | 8.2 | 331 | 17.4 |
| | | 476 | 19.1 | 33.2 | 17.9 | 28.6 | 30.6 | 7.6 | 232 | 12.2 |
| | | 419 | 19.0 | 33.9 | 17.8 | 29.7 | 29.0 | 6.6 | 192 | 10.1 |
| | 29y | 1300 | 24.6 | 31.6 | 23.4 | 25.5 | 66.6 | 8.9 | 592 | 20.4 |
| | | 748 | 25.4 | 37.2 | 25.4 | 31.5 | 58.3 | 10.2 | 592 | 20.4 |
| | | 419 | 24.3 | 43.5 | 24.2 | 36.6 | 44.0 | 8.8 | 386 | 13.3 |
| Gwent | 21y | 2057 | 23.1 | 20.3 | 20.7 | 15.9 | 41.0 | 9.4 | 383 | 18.4 |
| | | 872 | 23.6 | 24.7 | 22.4 | 21.2 | 30.8 | 9.8 | 301 | 14.5 |
| | | 872 | 24.0 | 25.7 | 23.7 | 22.6 | 34.9 | 10.2 | 356 | 16.9 |
| | | 478 | 24.5 | 29.0 | 23.7 | 26.6 | 26.6 | 10.1 | 270 | 12.8 |
| | | 459 | 23.8 | 29.5 | 22.9 | 26.3 | 25.0 | 9.6 | 241 | 11.5 |
| | 27y | 1996 | 27.5 | 22.0 | 22.9 | 17.1 | 45.9 | 9.5 | 437 | 16.4 |
| | | 872 | 26.5 | 27.5 | 25.6 | 24.3 | 40.4 | 10.9 | 440 | 16.3 |
| | | 478 | 26.7 | 32.5 | 26.2 | 29.8 | 33.3 | 10.9 | 363 | 13.4 |
| | | 459 | 26.6 | 32.6 | 25.9 | 29.3 | 31.0 | 10.5 | 325 | 12.0 |

**Table 7.2.** Average yields in tonnes dry matter $ha^{-1}\ y^{-1}$ of *Populus trichocarpa* × *deltoides* (RAP) in Forestry Commission trials* of short rotation coppice on three sites harvested in winter 1985/86.

| Site description | Alice Holt 2 year rotation† | Mepal 2 year rotation | Long Ashton 2 year rotation | Long Ashton 4 year rotation |
|---|---|---|---|---|
| Spacing | Wet heavy clay strong weed growth | Highly fertile fen peat | Shallow brown earth calcareous in parts | |
| 1 m × 1 m | 8.90 | 15.16 | 10.44 | 8.84 |
| 2 m × 2 m | 2.02 | 9.28 | 9.19 | 5.99 |

* This work was done under contract to the Department of Energy.
† The 2 m × 2 m result is unusually low due to heavy weed competition and the low number of coppice shoots per stool.

close spacings are not required for silvicultural purposes. Thus it is quite feasible to choose the spacing that meets any desired objective of maximising either individual tree growth, or volume per hectare or a particular age of maximum annual increment, etc. The choice is very much with the grower. Overall the following can be recommended.

1. Growing veneer butts to 40 cm d.b.h. (minimum top diameter 25 cm) on rotation of approximately 25 years.
   - 6 to 8 m spacing for *P. trichocarpa* cultivars and hybrids.
   - 7.5 to 8 m spacing for *P.* × *euramericana* cultivars.
2. Growing pulp wood or other fibre to maximise yield per hectare of utilisable stems (10–20 cm d.b.h.) on rotations of 10–15 years at 2–3 m spacing. This is particularly appropriate for clones or hybrids of *P. trichocarpa* which are more tolerant of competition at close spacing.
3. Growing biomass of stick size material on coppice rotations of 1–5 years at 1–2 m spacing.

## *Thinning*

The role of thinning in forestry is to provide the benefit of increased growing space to the remaining trees and to improve the quality of the stand by removing defective or misshapen trees as soon as possible and to provide intermediate financial returns. In traditional poplar cultivation neither of these benefits are needed: wide initial spacing substitutes or replaces the need for thinning in the life of the crop and the generally good stem form of poplars obviates the need to thin to favour the best stems. Of course thinning does provide an early return in conventional forestry but the short rotations used in poplar cultivation greatly diminish the importance of this possible consideration.

Even at close spacings thinning is generally dispensed with and rotation age used as the tool to control final tree size. As a rule, therefore, thinning plays little part in poplar silviculture.

## *Pruning*

In traditional poplar cultivation high pruning has been an important operation. The object is to eliminate side branches on the lower bole early in the life of a tree. This prevents formation of knots in the wood except near the core.

It is general practice to prune in two or three lifts to achieve a pruned bole of 6–7.5 m. However poplars generally develop a distinct whorl of branches at the start of each year's growth and pruning can with advantage be related to this fact. To produce the maximum volume of knot free timber and limit the diameter of pruned branches to 5 cm or less an early start to pruning is essential. With fast growing poplars the lowest whorl of branches and the small branches up to the base of the next whorl should be removed not later than the end of the third growing season. Thereafter it is recommended that no more than one whorl plus the interwhorl branches should be removed in any year so that pruning progressively lifts the height of the pruned stem to approximately 7.5 m within the first 10 years. During this period there must be 1 or 2 years without pruning to allow sufficient crown to develop commensurate with the increasing height of the tree. A diagram illustrating the pruning regime adopted by Bryant and May (Forestry) Ltd for plantation grown poplar is shown in Figure 7.1. Above 7.5 m the relatively small additional volume of knot free timber obtained by further pruning may not justify the cost on a normal 25-year rotation.

For most poplars once the main pruning has been done to cut off side branches further pruning of the stem is necessary to prevent epicormic branches developing. Control of such epicormics must be done frequently – at least every other year – but there is circumstantial evidence to suggest that if epicormic buds or young tender shoots are rubbed-off in June each year, over time new growth of epicormics tends to diminish. Clearly if such intensive treatment can be given all risk of pin knot formation from epicormics will also be eliminated.

### Pruning practices and techniques

Apart from treatment of epicormics in June pruning is best carried out in the winter months well before time of flushing. Generally poplars do not exude large quantities of sap or resin from wounds and there is little risk of infection by silverleaf disease.

Branches should be cut off before their diameter exceeds 5 cm at the base and preferably well before this diameter is reached. It is essential that only live branches are pruned, i.e. green pruning. Pruning chisels, long armed pruners and pruning saws on poles are all suitable tools.

Poplars generally exhibit excellent stem form – straight and cylindrical boles – and there is usually no need to single leading shoots or carry out pruning to shape the crown. Such operations are only practical during the first and possibly the second year of growth while the leading shoot is still within reach.

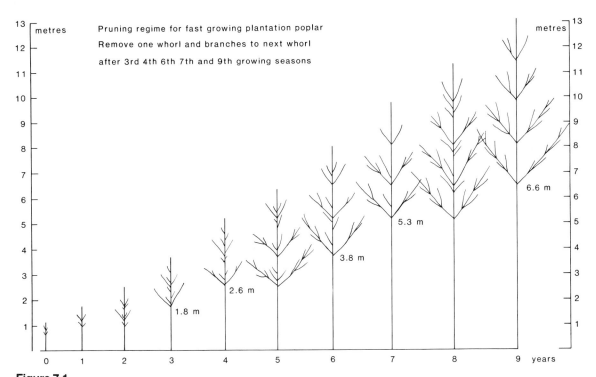

Figure 7.1

# Chapter 8
# Rate of Growth and Yield

## Rate of growth

The fastest growth rates of poplar occur in the main in sheltered, fertile nurseries. In southern Europe stems over 5 m long can be produced in a single season from hardwood cuttings, while two-year-old plants can easily reach a height of 8 m. In Britain, growth of this magnitude is never remotely approached even in well cultivated and heavily manured beds. Yet during the propagation stage poplars grow faster outdoors than any other plants. In the best nurseries in the south of England, most of the well-known cultivars raised for planting for wood production, screening or shelter can grow as much as 1.5–2 m in the first season. In warm summers some plants may be around 2.5 m tall. Established stools, cut back annually during the winter to supply cuttings or sets, may have even longer shoots than this. Selected cultivars of *P. trichocarpa* and some of its hybrids are usually the most vigorous and can have shoots 2.5–3.5 m long.

The record growth from a cutting in one season in this country is held by a hybrid bred artificially in the United States as long ago as 1924. It was given the patented name 10 years later of 'McKee Poplar'. Its parents were *P. deltoides* ssp. *angulata* and *P. trichocarpa*. Imported by the Forestry Commission in 1948 it soon proved to be an outstandingly vigorous clone. The tallest plants were grown in a small nursery at Alice Holt Lodge. Often increasing in length by more than 30 cm a week, they reached a height of 4 m in four months. Soon afterwards, 'McKee' was found to be highly susceptible to bacterial canker and plans to obtain its release to horticulture had to be shelved.

The large seasonal height increases commonly seen in the nursery are not often achieved in the field. Where they are, the vigour is seldom sustained in consecutive seasons. None the less there are well documented records of favourably located trees in Britain growing more than 2 m a year after planting out. The most distinguished performance is almost certainly that of *P.* 'Androscoggin' (*P. trichocarpa* × *maximowiczii*) and a clone of *P. trichocarpa* imported from Canada. Both were included in a trial of clones planted at Quantock Forest, Somerset in 1950. In the first season the most vigorous specimens of each grew 2 m. They then continued to grow at this rate until reaching a height of more than 25 m at the end of the 12th season. Remarkably, some of the trees remained vigorous for several more seasons to reach a notable 100 ft (30.5 m) only 18 years after planting. The tallest specimen of *P. trichocarpa* then had a breast height diameter of 44.5 cm while the biggest *P.* 'Androscoggin', though not quite 30 m tall, had a diameter of 45.5 cm.

This exceptionally fast rate of growth enabled strength and quality tests to be carried out on veneers and match splints at a much earlier age than usual. The youngest trees felled and rotary peeled were only 11 years old; their breast height diameters ranged from 27 to 30 cm. Subsequently both clones were found to be unsuitable for cultivation for veneer logs.

Though some of the well-known poplars in general cultivation in Britain may occasionally grow 2 m in a very warm summer, the largest annual height increases that can be expected as a rule over several seasons are around 1.7–1.8 m. Even then, this growth rate is possible only on the most fertile, sheltered lowland sites and it is unlikely to be sustained for more than the first 8 or 9 years. Thereafter vigour may be expected to decline gradually. But by the 25th season, when height increases may be less than 1 m a year, the trees should be more than 30 m tall. More often, though, poplars grow 1–1.3 m a year for the first 10–15 years and then by 0.6–1 m annually until 22–25 m tall at about 25 years. On the poorest poplar sites, the yearly height increase may be as little as 0.8 m to begin with, declining to 0.5 m annually by the 20th season. A height of 25 m may never be attained. Where growth is slower than this an infertile soil, poor drainage or exposure may be to blame.

On all but the most difficult sites, most of the commonly planted poplars, including many of the slower growing species and cultivars, can easily reach a height of 30 m or may be expected to do so with increase in age. Only a few white poplars and aspens, and certain species of the sections *Leucoides* and *Tacamahaca* have so far failed to attain this height and perhaps never will. It is probable in a few instances that favourably located specimens of some might have reached 30 m but for a serious attack by a debilitating disease.

The largest dimensions likely to be achieved by poplars in this country can only be guessed at. Measurements of 'champion trees' reported recently by Mitchell and Hallett (1985) provide a clue, however. They list five poplars 40 m or more in height which, hardly surprisingly, are amongst the tallest broadleaved trees in the British Isles. One, a fine specimen of 'Serotina' at Bowood House Estate, Wiltshire, is in fact the second tallest broadleaved tree recorded. When last measured in 1984 it had a height of 46 m and a stem diameter of 146 cm at 1.5 m. The fourth tallest broadleaved tree in their list is a specimen of 'Eugenei' at Colesbourne Estate, Gloucestershire; in 1984 it was 43 m tall and had a diameter of 135 cm.

Measurements given by Mitchell and Hallett (1985) of the tallest specimen of *P. trichocarpa*, also at Colesbourne, provide the clearest indication of the long-term performance and potential size of this important species and its related hybrids. The earliest introduction of this North American tree in 1892 proved to be susceptible to bacterial canker. While it is well known that several plantings of the clone grew well for a time, and that some trees reached utilisable dimensions, there are no records of any attaining an impressively large size. The Colesbourne tree, raised later from material imported in 1903, appears, on the other hand, to have remained free of serious bacterial infection. Last measured in 1984 when only about 80 years old, it was already 41 m tall with a stem diameter of 100 cm.

Whether the canker resistant cultivars of *P. trichocarpa* imported since, and only recently released after careful disease and field trials, will grow equally successfully to reach more than 40 m remains to be seen. Certainly their early growth rates are usually better than that of *P.* × *euramericana* cultivars, especially on poorer sites, and it is encouraging to note that they often remain more vigorous as they approach maturity. In its natural habitat *P. trichocarpa* has reached heights of 60 m (Sargent, 1905). Collingwood and Brush (1947) suggest that on sheltered sites near the coast it can be 175 to 225 ft (53 to 69 m) tall.

## Yield

For more than 30 years, since being made eligible for Forestry Commission planting grants, poplars have been planted at wide spacing for the production of veneer logs and sawlogs. To begin with the recommended spacing was 18 × 18 ft (5.5 m). Then in 1970 changes were made to the regulations, so that the maximum planting distance in plantations attracting grant aid became 8 × 8 m. In 1988 a new grant scheme was introduced and poplars were not differentiated from other broadleaved species, though grants will only be paid on approved clones. At this spacing it was and still is expected that most trees in the stand may reach veneer log dimensions without the need for a thinning. The rate of growth and volume production of 20 trees including poplar in stands at wide spacing have been well documented by the Forestry Commission in Imperial and metric yield tables. The construction and presentation of the metric tables and the provision of advice on their use was undertaken by Hamilton and Christie (1971).

Poplar volume data along with other information on growth are shown in six tables. The tables are basically a series of growth rate categories known as yield classes[1] which reflect the range of maximum mean annual volume increment[2] commonly encountered in British conditions. The largest maximum mean annual increment shown for poplar is 14 m$^3$ ha$^{-1}$, that is yield class 14, while the lowest is 4 m$^3$ ha$^{-1}$, that is yield class 4. The intermediate yield classes are 12, 10, 8 and 6.

Each of the six yield class tables includes top height,[3] mean diameter and basal area data as well as stand volumes in cubic metres to top diameters of 7 cm (cumulative production), 18 cm and 24 cm. Values are shown in each case at 5-year intervals. All the data in the tables relate to stands planted at a square spacing of 7.3 m (24 ft), that is 185 trees per ha, in which no thinnings are carried out.

Yield class 14 stands are found in Britain on fertile soils in sheltered mild localities. They are mainly in the southern half of England. Most trees are large enough for conversion into veneer logs at about 25 years of age, when the total crop volume to a top diameter of 7 cm is likely to be around 320 m$^3$ ha$^{-1}$. About 90 per cent of the tree should then be large enough for rotary peeling, though the amount of veneer quality wood available will obviously depend on the height to which pruning was taken, the prevalence of epicormic shoots and the arrangement, size and number of branches in the crown.

Many stands in Britain are, of course, better than yield class 14 but, as a rule, they are confined to the best farm land and to well drained water meadows and carr. Where prime sites can be planted, crops of *P. × euramericana* cultivars such as 'Robusta' are able to reach veneer log specifications on rotations as short as 22 years. The majority of stands of *P. × euramericana* tend to be yield class 8 to 12, however, and rotations of 30 years or longer are commonplace. On the poorest sites (yield classes 4 and 6) black poplar hybrids grow despairingly slowly and expectations of a crop of veneer logs may not be fulfilled, even on greatly extended rotations.

Yield forecasts are based almost wholly on the performance of the well-known, older euramerican hybrids that have been in cultivation in this country since the turn of the century. However, they are intolerant and site demanding poplars. But in clonal trials established during the past 30 years or so, and in a few commercial crops planted only in the past few years, several Balsam poplar clones have been found to be appreciably faster growing and much more productive and accommodating than any black poplar hybrid. Almost certainly the best example is provided by the carefully recorded behaviour of balsam poplars in a clonal trial planted in Quantock Forest, Somerset, in 1950. The exceptionally fast rate of growth of the two most vigorous clones – *P.* 'Androscoggin' and a Canadian provenance of *P. trichocarpa* – has already been commented on in 'Rate of growth' on page 58. Their volume production, assessed at intervals during the life of the trial, was no less remarkable. At 18 years of age both clones were found to have a total volume production of 346 m$^3$ ha$^{-1}$ compared with 232 m$^3$ for *P.* 'Robusta', the fastest growing black poplar hybrid. Planted at a stocking of 331 trees per ha, all three clones had a stocking of 269 trees per ha at time of assessment. Both the balsam poplar clones were classified subsequently as yield class 20+.

Unfortunately, excepting three satisfactorily disease resistant cultivars, these and other vigorous balsam poplar clones have proved to be too susceptible to bacterial canker to be recommended for general planting. Yet all the evidence indicates that prospects of higher yields rest almost wholly on the use of selected cultivars of *P. trichocarpa* and its hybrids, especially *P. × interamericana (P. deltoides × trichocarpa)*. Probably for the foreseeable future, cultivars of

---

1 Yield class: A classification of rate of growth in terms of the potential maximum mean annual increment per hectare of volume to 7 cm top diameter, irrespective of age of culmination or tree species.
2 Mean annual volume increment (MAI): The total volume production to date, divided by the age, that is the average rate of volume production over the life of the crop to date. The maximum MAI reflects the maximum average rate of volume increment which can be achieved on a particular site.
3 Top height: The average total height of the 100 trees of largest diameter per hectare.

these poplars including, in particular, clones bred at and recently released by the Poplar Research Station in Belgium offer the best hope of achieving consistently high yields of veneer logs and sawlogs.

It seems likely, too, that the same balsam poplar clones may be the best available for growing on short rotations at close spacing. In the Near and Middle East, white and black poplars have been cultivated for centuries in closely planted stands to produce wood – usually of small dimensions – for domestic and farm use. Planting at high density has been adopted locally by farmers in southern Europe as well but the practice has not yet spread to Britain or other countries in northern Europe.

Volume assessments in a poplar spacing experiment in Alice Holt Forest has shown that pulpwood yields of *P. trichocarpa* at close spacing are greater up to 30 years of age than the yields normally attainable by other trees. The data given in Table 8.1 are for trees planted at spacings of 2.7 × 2.7 m, 3.6 × 3.6 m and 4.6 × 4.6 m in plots left unthinned. Though the initial object of the experiment was to study the effect of initial tree spacing on pulpwood volume production, its duration was extended at 11 years of age to determine the pattern of cumulative volume production for different initial spacings without interference from thinning. The experiment was to have been felled when 12 years old to study coppice production.

It is evident from Table 8.1 that as spacing is increased, mean breast height diameter is increased but, at least up to 22 years of age, yield per hectare decreases. Height increment is not affected by spacing. The data also confirm that *P. trichocarpa* is able to maintain high growth rates after canopy closure, an attribute not shared by other poplars.

Provisional profitability assessments for close spaced balsam poplars grown for pulpwood have been made by Stern (1972). His figures are based on the behaviour of *P.* 'Balsam Spire' at a square spacing of 2.2 m. This artificial hybrid (*P. trichocarpa × balsamifera*) has a somewhat slower rate of radical growth than the recently released cultivars of *P. trichocarpa* and *P. × interamericana* and, as a consequence, its yield falls below the figures shown in Table 8.1 for a given spacing. Because it has a highly fastigiate form, however, it can be cultivated at slightly narrower spacings than might ordinarily be adopted. Stern also includes profitability assessments for poplar coppice, suggesting increased returns where two 15-year coppice rotations follow a 12-year maiden crop, but his calculations are not supported by yield data and the conclusions are tentative.

Estimates of yields of poplar biomass produced on mini-rotations, usually of 1–8 years, are based wholly on experimental evidence. The high number of variables being tested in experiments inhibits conclusions being reached

**Table 8.1.** Height, diameter and volume of *P. trichocarpa* between 9 and 29 years (Alice Holt Forest)

| | Spacing (m) | | | | | | | | |
|---|---|---|---|---|---|---|---|---|---|
| | 2.7 × 2.7 | | | 3.6 × 3.6 | | | 4.6 × 4.6 | | |
| Age (yrs) | Mean height (m) | Mean diam. (cm) | Volume per ha ($m^3$) | Mean height (m) | Mean diam. (cm) | Volume per ha ($m^3$) | Mean height (m) | Mean diam. (cm) | Volume per ha ($m^3$) |
| 9 | 11.1 | 14.9 | 94 | 11.5 | 17.4 | 69 | 11.2 | 18.4 | 48 |
| 11 | 13.8 | 16.5 | 155 | 14.5 | 20.2 | 131 | 14.0 | 22.5 | 94 |
| 13 | 16.6 | 18.3 | 229 | 17.1 | 22.6 | 199 | 16.4 | 25.7 | 152 |
| 15 | 16.1 | 19.4 | 271 | 17.0 | 24.1 | 232 | 16.2 | 27.6 | 172 |
| 19 | 18.9 | 21.6 | 366 | 19.7 | 26.8 | 339 | 19.1 | 30.8 | 247 |
| 22 | 19.6 | 22.9 | 421 | 21.6 | 28.4 | 402 | 20.9 | 32.7 | 298 |
| 29 | 23.4 | 25.9 | 607 | 25.3 | 32.0 | 621 | 25.1 | 37.3 | 461 |
| | (Mean stocking 1275 trees per ha) | | | (Mean stocking 748 trees per ha) | | | (Mean stocking 449 trees per ha) | | |

quickly on either practice or productivity. For example, more than 70 clones of poplar are listed by the International Energy Agency in an inventory of species and cultivars potentially valuable for forest biomass production published in 1981 by the National Swedish Board for Energy Source Development. Similarly, stockings of plants in experiments vary enormously. Cannell (1980), using a systematic fan-shaped 'Nelder' design, examined 25 stockings ranging from 500 plants per ha up to 500 000 plants per ha. In 'Nelder' experiments planted by the Forestry Commission in 1981 and 82, 11 stockings (1322 to 160 000 plants per ha) were compared. Where large blocks of plants have been established in the field only two or three stockings have been included because of the difficulty of finding large, uniform sites for the experiments. Two stockings (2500 and 10 000 plants per ha) are being compared in Forestry Commission experiments. Other highly variable factors include length of rotation and soil type and climate. In Britain oven dry yields of poplar biomass in 5 and 6-year-old field experiments have ranged from 7 to 10 t $ha^{-1}$ $y^{-1}$, while in old, vigorous stoolbeds, yields of 20 t $ha^{-1}$ $y^{-1}$ have been obtained (see Table 7.2). The highest yields have been obtained with the artificial hybrid P. 'Rap' (*P. trichocarpa* × *deltoides*) and selected clones of *P. trichocarpa*.

# Chapter 9
# Poplars and Farming

The integration of poplar cultivation with agriculture is a long established practice over much of western and central Europe and extends into the Middle East. In Great Britain, although the grazing of sheep and cattle in older plantations is not uncommon, arable intercropping on any scale seems only to have been practised by the match industry during the 1960s and 1970s. Poplars may often be planted on land suitable for the species which, because of winter flooding or difficulty in regular and timely cultivation, is otherwise unsuitable for agricultural crops. Provided land which is liable to flooding is adequately drained and the period of flooding is of short duration, not exceeding a few days in the growing season or a week or so in the winter, the cultivation of poplars may give a useful financial return as well as contributing to flood control. If the land is suitable for the summer grazing of sheep and cattle the shade and shelter provided by well established, widely spaced, poplars can be beneficial. Since their rotation is relatively short compared with other tree species poplars offer a useful short-term alternative crop for a proportion of farmland when, because of economic or political factors, conditions are less encouraging for the production of food crops. In Italy the returns from growing poplars on short rotations of about 15 years are carefully balanced against the returns from other crops and land may be used for poplar cultivation in rotation and in mixtures with other crops and with grazing, producing returns from sales of timber which may more than compensate for the reduced production of other crops. Integration of poplars with farming may also utilise labour and farm machinery at a time when there is little other work on the farm. In parts of southern Europe more than 20 per cent of land may be utilised for poplars. The cultivation and tending of other crops generally encourages fast growth of poplars although the yields of crops cultivated with poplars are between 5 per cent and 25 per cent less than would be obtained outside the plantation. The range of arable crops grown in association with poplar plantations includes wheat, rice, maize and other cereals, pulse crops, cotton and melons.

## *Line planting*

Poplars are frequently planted as single or double lines either for wood production in areas poor in timber or to provide shelter as windbreaks. In southern Europe lines may be spaced 100 m or even 200 m apart to provide useful periodic income from sales of timber while reducing to a minimum the ground taken out of agriculture. In Great Britain poplars, particularly early flushing varieties such as 'Balsam Spire', are frequently planted in single lines at spacings as close as 1.5 m so as to provide protection from cold winds. Such protection is particularly valuable for hops, fruit orchards and other horticultural crops and, in parts of East Anglia, lines of poplar may reduce wind erosion of finely cultivated peat soil in the springtime.

## *Arable cropping*

In England during the 1960s and 1970s Bryant and May developed successful arable intercropping practices on their estates in East Anglia and Herefordshire. The prime objective was the production of veneer quality timber for match splints. Trees were planted at 185 trees per hectare and the development of intercropping arose from the desirability of making full use of agricultural land, particularly during the first few years after planting. Accordingly the intercropping methods adopted were modifications of normal arable practice designed to meet the overriding objective of veneer timber production. For farmers seeking alternatives to food production the integration of poplars with agriculture offers a range of options and timber production may be only one of several objectives.

The intercropping practices developed by Bryant and May are described by Miller (1976) and Beaton (1987). The poplars were generally planted on a triangular pattern at a spacing of 7.9 m between trees giving distances between rows of approximately 6.8 m. During the first growing season plantations were repeatedly cultivated, maintaining clean fallow conditions so as to promote the maximum growth of the poplars. To have continued clean fallow cultivation in this way during the first 10 years or so would have ensured the maximum rate of growth of the poplars. However in order to utilise land more effectively in the first 7 or 8 years a system of arable intercropping was practised from the end of the first growing season. There was sufficient space between the tree rows to sow one drill width centrally in each bay and maintain clean fallow strips approximately 1.8 m wide on either side of the tree rows, or to sow two drill widths in every bay without any fallow. In practice it was found that the single drill interow cropping with adjacent fallow strips caused only a slight reduction in growth rate of the poplars but that double drill cropping caused a severe and unacceptable reduction in growth of poplars. The single drill arrangement was thus practised, allowing arable cropping of one-third of the gross field area to continue for 7 or 8 years. However difficulties arose from the repeated cropping of the same strips of land; diseases increased, particularly 'take-all' (*Ophiobolus graminis*) in cereals, while the increasing shade cast by the poplars from the third season onwards delayed the ripening and harvesting of arable crops, and consequently led to delays in autumn sowing. It was also necessary to carry out pruning of the poplars and to dispose of the pruned branches by chopping these on the arable strips following harvest. These practical problems were very largely resolved by the development of an alternate bay system with two drill widths cropped in every other bay and the intervening bays maintained as clean fallow during the growing season. Following harvest the stubble bays were left so as to accommodate poplar tending operations during the winter, before being cultivated to clean fallow, while the summer fallow strips were sown to arable crops each autumn thus giving in effect a two-course rotation of crop and fallow in every bay. The total area cropped in this way remained at one-third of the gross field area. Both these systems gave useful returns from cropping up to the 7th or 8th year.

The importance of fallow cultivation on the growth of poplars under either system was well demonstrated by trials on the Herefordshire estates comparing the effect of cropping one-third and two-thirds of the gross field acreage against a control treatment of clean fallow throughout the growing season. For *P*. 'Robusta', trees in the control plots maintained the fastest growth rate; for trees with one-third intercropping, with fallow strips on both sides of the tree lines, the rate of growth was less but still acceptable; but for trees with two-thirds intercropping and consequently no adjacent fallow, shoot growth was reduced to an unacceptable level during the first few years following planting and to as little as 10 cms or less in the second year. Similar trials on the fen peat soils of East Anglia, high in available nitrogen, gave quite different results and trees with one-third and two-thirds of the gross field area under wheat showed little difference in growth rate to those in control. These observations indicate that for 'Robusta' and probably for most *P*. × *euramericana* varieties fallow cultivations during the first few years can be of critical importance in early establishment particularly where the available moisture and nutrients may be limiting in some seasons, and intercropping practices should take this factor into account. *P. trichocarpa* and hybrids with this species are generally more tolerant of competition and may respond differently to intercropping practices. The trend for larger farm machinery in the last decade or so may require some adjustment to the spacing and arrangement of poplars when intercropping is intended. The working width of seed drills and combine harvesters can be used to determine a modular width of cropping sufficient to allow machinery to operate at a safe distance from the tree rows and to turn without difficulty at the headlands. The distance between trees within the rows can then be calculated so as to give the desired stocking.

The period during which arable intercropping will give satisfactory yields in poplar plantations will be influenced by the width between rows and the rate of growth of the poplars but is unlikely to extend for longer than about 10 years. As the poplars grow the increase in shade and possibly the increased competition from tree roots will cause a progressive decline in yield to the point where further arable intercropping becomes uneconomic.

## *Grazing*

When poplars reach a sufficient size to withstand the attentions of cattle and sheep it is often beneficial to allow grazing within the plantation. Following the end of the arable intercropping phase on the Bryant and May Herefordshire estates, usually between the 8th and 10th year, plantations were systematically sown to a permanent grass/clover ley, fenced, watered and let for summer grazing. At that age the *P. × euramericana* cultivars had a sufficiently rugged, woody bark to withstand the occasional attempt by sheep and small cattle to chew the bark. *P. trichocarpa* cultivars and the 'Balsam Spire' hybrid still have a relatively thin bark at that age and these varieties were frequently damaged by sheep, the animals apparently finding the cambium and associated tissue attractive to eat. Damage can be so severe that trees are girdled and killed and crops can become very seriously damaged in a day or so. The presence of rams in a flock of sheep or of horses can be disastrous for the tree crop and these animals must never be allowed in poplar plantations at any age. Under most conditions grazing in plantations is best restricted to the 6–7 months from early April to the end of October with stock being removed from plantations before the first autumn frosts. Normal prudence in stock control particularly in relation to the intensity of grazing in plantations is essential if damage to trees or pastures is to be avoided. The quality of the grass ley will decline as a direct result of the progressive increase in shade and in autumn leaf fall, the rate of decline being influenced by the density and vigour of the tree crop and the variety of poplar. The timing of flushing in the spring, of leaf fall in the autumn and possibly differences in leaf structure between varieties may all influence the blanketing effect of the winter leaf mulch. In some situations useful grazing may continue until the poplars are mature and ready to be felled.

Grazing in young poplar plantations during establishment cannot be contemplated without adequate physical protection to the young trees. Plastic guards with two stout stakes in support and providing a complete barrier up to 1.5 m in height may give sufficient protection against sheep from the first year onwards. Cattle however cannot be introduced into young plantations and are probably best excluded until the trees are able to withstand their presence without protection.

# Chapter 10
# Characteristics, Properties and Uses of Poplar Wood

## Wood characteristics

The timbers of the various species and cultivars of poplar grown in Britain have certain essential features in common. However, the quality of the wood varies considerably according to the site conditions, habit of growth and the attention given during cultivation.

Aspen, the black poplars including *P. × euramericana* hybrids and the balsams have uniformly white, yellowish-white, pale brown or greyish wood when freshly felled. The heartwood and sapwood are usually the same colour, though the former is sometimes slightly darker in shade. Only the white poplars and grey poplar show a contrast between the whitish coloured sapwood and the pink or reddish heartwood. Freshly cut wood usually has an unpleasant smell. On drying out the colour of the wood fades but may never become white. The heartwood of the euramerican hybrids sometimes show brown, green or reddish streaks. Dried wood is virtually odourless.

The anatomical features for all the species and cultivars are so similar that a single description will suffice. On transverse sections annual rings are fairly distinct, their boundary marked by a narrow band of storage parenchyma cells and a zone of denser summer wood contrasting with the succeeding broad zone of coarse textured spring wood. There is no difference in colour between spring wood and summer wood. On longitudinal sections the rings are not easy to differentiate and lines of vessels are seen as fine scratches. Rays are very numerous and extremely fine, just visible to the naked eye on radial and quarter-sawn surfaces as silvery lines but not clear on end-surfaces even with a lens. The grain is usually straight and even, and the texture is generally fine and even due to the small size of the fibres, uniform structure and lack of contrast between spring wood and summer wood. Pith flecks, commonly due to insect attack, are sometimes present, and small pin knots, associated with the occlusion of epicormic shoots or presence of dormant buds may be seen as well.

The wood of poplar is diffuse-porous, with hardly any difference in vessel number and size in the spring wood and summer wood. The fibres, which account for half to three-quarters of the woody tissue, are relatively long and thick-walled. Descriptions of the vessels, fibres and parenchyma cells in medullary rays are well documented (FAO, 1979).

## Wood properties

Poplar is one of the lighter and softer home-grown timbers but its strength properties, especially toughness, are relatively high when its density is taken into consideration. The euramerican cultivars have the following densities (FAO, 1979):

- density of green wood (green weight/green volume) varies between 700 and 1050 kg m$^{-3}$;
- density of air-dried wood (weight/volume at 12 per cent moisture content) varies between 300 and 550 kg m$^{-3}$;
- density of oven-dried wood (dry weight/green volume) is usually between 320 and 440 kg m$^{-3}$ (range is 280 to 520 kg m$^{-3}$).

The strength properties of poplar fall between those of European whitewood and British-grown Norway spruce. Bending strength, stiffness, impact, compression and hardiness values reported by Lavers (1969) are shown in Table 10.1.

The moisture content of freshly felled poplar wood is nearly always high, though season of sampling and zone of sampling in the tree can greatly affect results. Big differences in moisture content also occur between sites and between species and cultivars. The green weight is usually at least twice the dry weight; it is never less than 75 per cent greater and, in exceptional circumstances, it may be 300 per cent greater than dry weight. The wood dries well and quickly, and shrinkage is quite small (the volume decrease is never more than 17 per cent and is usually between 11 and 13 per cent). Longitudinal (axial) shrinkage is about 0.2–0.6 per cent, radial shrinkage is in the region of 2 per cent and tangential shrinkage is about 5.5 per cent (Farmer, 1972). Local pockets of moisture are apt to remain in the timber.

Moisture content at felling can affect the extent and type of shake, the extraction, transport and storage of round timber, the sawing and peeling properties of logs, and the conversion and drying of sawnwood and veneers. The presence of tension wood in logs can increase seasoning problems because of the risk of bowing of planks and buckling of veneers.

Poplar woods are inclined to be woolly in sawing and occasionally bind on saws, while some may be difficult unless cutters with thin sharp edges and reduced cutting angles are used.

The wood usually glues well, takes stain readily though sometimes with rather patchy results, and takes paint, varnish and polish satisfactorily. It is easily dented but does not readily splinter, and there is a tendency for small holes to close up. Bruising rather than splintering is a useful feature when the wood is subjected to abrasion. The wood burns quickly, with little resistance to flame penetration and it has, in general, low fire-resistance properties. It smoulders instead of igniting when violent friction is applied.

## Uses of wood

Although the wood of poplar superficially resembles coniferous timber, and has on occasions been regarded as a softwood substitute, it is valued as a special purpose wood with almost unique properties. As with all special purpose woods, however, a use does not necessarily imply the existence of a market.

Seasoned wood is liable to attack by common furniture beetle (*Anobium punctatum*), but it is immune to powder post beetle (*Lyctus brunneus*) attack. It is classed as perishable, being very liable to decay under conditions favourable to

Table 10.1. Strength properties of wood of *Populus × canescens, P.* 'Serotina', *Picea abies* and *Pinus sylvestris* at 12 per cent moisture content

| Species | Nominal specific gravity | Maximum bending strength<br>Modulus of rupture (N/mm$^2$) | Stiffness<br>Modulus of elasticity (N/mm$^2$) | Impact<br>Maximum drop of hammer (m) | Compression<br>Maximum compression strength parallel to grain (N/mm$^2$) | Hardiness<br>Resistance to indentation on side grain (N/mm$^2$) |
|---|---|---|---|---|---|---|
| European whitewood |  |  |  |  |  |  |
| *Picea abies* | 0.38 | 72 | 10200 | 0.58 | 36.5 | 2140 |
| *Populus* 'Serotina' | 0.38 | 72 | 8600 | 0.56 | 37.4 | 2220 |
| *P. × canescens* | 0.43 | 76 | 9500 | 0.61 | 36.9 | 2360 |
| Norway spruce (UK) |  |  |  |  |  |  |
| *Picea abies* | 0.35 | 66 | 8500 | 0.51 | 34.8 | 2000 |
| Scots pine (UK) |  |  |  |  |  |  |
| *Pinus sylvestris* | 0.46 | 89 | 10000 | 0.71 | 47.4 | 2980 |

fungal attack. In common with most timbers the sapwood of poplar is readily impregnated with preservatives. As a consequence thinnings, which contain a high proportion of permeable sapwood, can be used as fence posts and stakes after preservative treatment. In contrast the heartwood is difficult to impregnate even under pressure and is best avoided for permanent estate work.

Rotary peeling of poplar to produce thin veneers can be undertaken without any pre-treatment of the logs. Match splints and match boxes are the best known products made from veneers. Only logs of high quality are accepted by the match trade. They must be fresh felled with bark intact, round and straight, free from knots, soft centres, splits, shakes, worm and insect damage and any other serious defects. The ends must be well trimmed and free from flutes. As a rule logs with a diameter less than 25 cm at the top end or greater than 60 cm at the butt end are not acceptable to the match industry. A diameter of about 38 cm is favoured. To ensure that splints have the required degree of flammability, they are dipped at one end into paraffin wax. In Britain neither imported nor home-grown poplar have been peeled by the match trade since 1978.

Another common use for rotary peeled poplar is in the manufacture of light packaging and small containers for the food industry. Some of the best known domestic wooden receptacles – boxes for cheese and confectionery such as Turkish delight and chip baskets for soft fruits – are made wholly from poplar. For this purpose poplar has certain outstanding advantages including its clean appearance, light weight, lack of odour and freedom from resinous or oily products likely to exude and contaminate the food. Moreover the peeled veneer can be bent at right angles and stapled without breaking. In this country containers made from poplar have given way in recent years to plastic and pulped equivalents and relatively little soft fruit is now retailed in wooden punnets.

Much poplar veneer is also used in the manufacture of both disposable and returnable crates and boxes for carrying hard fruit and vegetables. A good deal of the farm and horticultural produce imported into Britain reaches our markets and shops in crates made almost entirely from poplar. At the present time there are comparatively small manufacturing outlets for vegetable crates in this country.

Poplar veneers are extensively used as well in the manufacture of plywood. In Europe the plywood is usually made with three plies and can be as thin as 3 mm. But poplar is also used in plywood up to 25 mm thick, when multiple plies are bonded together. Poplar is liked because of its light colour and good peeling, drying, glueing, sanding and finishing properties. Glues commonly in use are acceptable though the plies may be discoloured by some types of dark glue, and surfaces can be readily stained, painted or varnished. A lot of the plywood finds its way into light furniture, high-quality packaging and the overlay of hollow doors. In Italy in particular, where there is a long history of marketing and utilising poplar timber, poplar veneers and plywood are used to surface relatively expensive furniture, often cupboards in kitchen units. Natural or coloured finishes are available. In Britain the employment of poplar in plywood manufacture tends to be small and sporadic.

Some packaging is made from sliced poplar. Logs are sliced green and veneers about 6 mm thick can be cut. Sawn poplar wood is used primarily for packaging in Europe. In Italy nearly all packaging wood is poplar, and in France poplar accounts for half the packaging output. Poplar is sought after because of its light weight, good mechanical properties, the absence of taste or smell and its ability to hold nails and staples. Much of the packaging made from poplar in central and southern Europe goes to fruit and vegetable growers. Often, the cultivation of poplars for the production of wood for packaging is closely integrated on farms with agricultural and horticultural cropping. Poplar packaging specifications have been prepared in recent years for a very wide range of boxes and crates, including nailed containers for sea transport, for both natural and manufactured products.

Sawn poplar wood is also used as supports and struts in packaging made from hardwoods, plywood, fibreboard, particle board and cardboard. In central and southern Europe much of the poplar utilised for packaging is of comparatively low quality. Low quality wood is also used in the manufacture of pallets.

High quality sawn wood is much used in furniture manufacture and, notably in the Near East, as a construction timber for framework and roofing. First-quality wood is also sought after by local craftsmen for interior joinery work for doors, shelving, casings and other domestic fittings. In Britain, sawn poplar is used as a mining timber and, decreasingly, to make brake blocks on account of its low flammability, and for wagon bottoms and wheelbarrows because of its non-splintery nature.

Throughout Europe and in many Asian countries poplar is used in the manufacture of wood wool, wood wool slab, pulp and paper, fibreboard and particle board (wood chipboard). The use of poplar in particle board reduces the weight of the board and gives it a pleasing clear colour. Particle board, used primarily in the construction and furniture industries, can be faced with veneers of poplar or other woods. In fibreboard manufacture, poplar can be used to replace other timbers even though the fibres of poplar are shorter than those of conifers. For neither product is debarking of logs required.

For many years poplar has been used to produce mechanical pulp for the manufacture of printing, writing and coated papers. Though less strong than mechanical spruce pulp, poplar pulp is whiter and has better opacity and absorbency properties. In recent years poplar has been used increasingly in the manufacture of semi-chemical pulp cooked with sodium monosulphite. The product has very high tensile strength and brightness properties.

Wood wool slabs made from poplar, used by the construction industry to provide heat and sound insulation, are the equal of or sometimes rather better than the equivalent conifer product. Shavings are also made into spills and, notably in southern Europe, into cheap and expendable holiday beach bags and hats by plaiting the strips.

In many parts of the world poplar leaves are fed to livestock as supplementary fodder. Though the chemical composition of the foliage varies from species to species, the commonly cultivated poplars such as *P. alba* and *P. nigra* have a nutritional value similar to that of grass and lucerne.

# Chapter 11
# Pests and Diseases

## Insect pests

Poplars are hosts to a large number of insect species. Only a few of these, however, are of any importance.

Possibly the commonest insects on these trees are the small poplar-leaf beetles, *Phyllodecta vitellinae* and *P. vulgatissima*, but even these are not important except when fairly large numbers are present. The adults are about 5 mm long and, in colour, respectively metallic green-gold and blue. The larvae are dirty grey-brown spotted with black, and when fully fed are about the same length as the adult. The immature stages feed side-by-side on the underside of the leaves, removing the lower leaf surface but leaving the veins. Such feeding causes the death of the adjacent tissue and characteristic irregular brown patches on the upper leaf surface. Damage by these beetles may be of importance in the stool bed, in the first year after planting out, or in newly bedded-out mist propagated plants. The loss in shoot growth after heavy attack in the stool-bed can be striking (particularly when clonal differences in susceptibility to attack and comparisons between affected and unaffected plants can be made). Fresh plantings are vulnerable for the first year or so, and severe damage may take place if neighbouring older poplars are infested. Stock raised by the mist-propagation method is usually of small dimension and may have at first flushing, as few as three leaves; damage should at this time obviously be kept at a minimum by insecticidal control.

The large poplar-leaf beetle (*Chrysomela populi*) may be particularly damaging to aspen. The adult beetle is brick-red and similar to a ladybird. The larva, which is about the same length as the adult (from 10–12 mm) is whitish with black spots and head. Both stages feed together and can completely strip young trees of their foliage, causing shoot dieback. The larvae at first feed on the surface tissues of the leaves, skeletonising them, but later feed, as do the adults, by removing chunks from the lamina. There may be two or even three generations a year.

Larvae of the white satin moth (*Leucoma salicis*) occasionally cause complete defoliation of large trees. The larvae (up to 45 mm long) are dark grey with a prominent white dorsal patch and bright red patch on each segment. They feed, often gregariously, from April to June.

The larvae of the poplar shoot borer (*Gypsonoma aceriana*) may cause failure of the leading shoot and subsequent forking in young crops, this being especially notable in 'Balsam Spire'. Damage occurs in May and June when larvae bore into the shoots and can be located by the brown tubes of excreta and silk in the leaf axils.

There are several other occasional pests which deserve some mention, and amongst them a number of attractively coloured sawflies and the puss moth (*Cerura vinula*) are perhaps the commonest. The leaves may also be mined by more than one kind of insect, and there are aphid species that attack leaves, shoots and even main stems.

During June to August two aphids produce characteristic galls on the leaf petioles, mainly on *P. nigra* and its varieties, also on *P.* × *berolinensis*, but not on North American poplars. The galls are of a spiral form when caused by *Pemphigus spyrothecae*; the round or pear-shaped galls are due to *P. bursarius*. The latter species also feeds on lettuce roots. However, the use of modern varieties of lettuce resistant to this aphid, together with the use of insecticides, means that this aphid is of less importance than formerly.

Defoliating insects such as leaf beetles and sawflies in stool beds can be controlled by spraying with an insecticide recommended against caterpillars. No specific advice can be given here; potential users are referred to Forestry Commission Booklet 52 for further information.

A number of wood-boring insects attack poplars. The poplar cambium midge (*Agromyza* sp.) larvae are opaque white, about 20 mm long but only about 1 mm wide, and bore long irregular tunnels in the cambium during the summer. These are packed with frass, and due to bacterial action quickly become brown. When the stem is cut transversely they appear as concentric rings of flecks; on logs sawn or split lengthwise they form narrower brown streaks.

The larvae of three moths occur quite commonly. They are the goat moth (*Cossus cossus*), which can be distinguished by its large size (up to 85 mm) and red-brown colour, the leopard moth (*Zeuzera pyrina*), and the hornet clearwing (*Sesia apiformis*); the last-named sometimes occurs in quantities on older trees, and can do serious damage. The large longhorn beetle (*Saperda populnea*), and the weevil *Cryptorhynchus lapathi*, both attack young trees, and are therefore more serious pests. In both cases it is the larva that bores into the poplar stem, and in the case of the longhorn a swelling and distortion of the stem or branch results. The only practical control measure against these two is the cutting-out and burning of infected material.

## *Diseases*

In Britain the worst affliction from which poplars cultivated for their timber suffer is bacterial canker caused by *Xanthomonas (Aplanobacter) populi*. The disease is first noticeable in the late spring as small slits in the bark of the twigs, from which a whitish bacterial slime exudes; later the bark dies around the slits. Very small twigs are often girdled and die, but on the main stem and branches cankers are formed which sometimes continue to develop for many years without causing girdling. In some cases very large complex cankers are produced by the combined effect of the necrosis caused by the bacteria and the healing efforts of the tree. Often, however, after a few years the canker encircles the stem, which then dies. Infection occurs principally in the spring through stipule scars, caused by the natural shedding of the stipules, and through bud scale scars. Infection also occurs in autumn through leaf scars but there is less inoculum present at that time. Poplars are seldom killed outright by this disease, but the death of large numbers of branches can cripple the growth of susceptible clones, while just one or two large stem cankers can spoil the trunk for veneer cutting.

The principal means of dealing with the disease is to plant cultivars which have been bred so as to combine resistance with other desirable qualities. The UNAL cultivars, developed in Belgium, are an outstanding example of this (see Appendix 2). Breeding for resistance only became possible after the discovery of the causal organism by Ridé (1958) and the subsequent development of standardised methods of artificial infection. Resistance has been one of the main features taken into account when lists of clones eligible for Forestry Commission planting grants are compiled. Even where these clones are used, however, it is recommended that no cankered trees be retained close to the plantations.

There are no other diseases nearly as common or as damaging in British poplar plantations as bacterial canker but a few others are sufficiently serious to warrant mention here.

The fungi *Marssonina brunnea* and several *Melampsora* species can occasionally cause premature defoliation serious enough to result in measurable loss of increment. The defoliation may also prevent the proper ripening of shoots and these may then be killed back by autumn or winter frosts. Leaves infected by *M. brunnea* (usually the more damaging of the two pathogens) are peppered with tiny dark brown spots and where infection is heavy, leaves quickly fall. The related fungus *M. populi-nigrae*, can attack Lombardy poplar severely, sometimes causing defoliation of all but the top of the crown. Repeated attacks over several years can kill entire trees (Peace, 1962). *Melampsora* species are typical rust fungi, alternating between the poplar and a completely different plant species during their annual life cycle. Thus *M. allii-populina* alternates with onion, and *M. larici-populina* with larch. In summer the infected poplar leaves are covered with bright yellow and orange spore pustules. The leaves wither and hang for a while on the tree before falling.

In Britain, our experience hitherto has been that symptoms appear late in the season, with the result that losses have involved only a small reduction in increment. However, much earlier defoliation has occurred on some clones on the Continent, leading to severe secondary attack by the canker fungus *Dothichiza populea* and even to death of entire trees. These severe outbreaks have been due to the emergence of races of *M. larici-populina* which are aggressive to some poplar clones that were resistant to the previously known race (Pinon, 1986).

Outbreaks of both diseases are heavily dependent upon the susceptibility of the poplars, the weather conditions during the growing season and, in the case of *Melampsora*, the proximity of the alternate host plants. Chemical control is never resorted to in plantations, but may sometimes be justified in nurseries.

*Melampsora* should not be confused with the relatively unimportant poplar leaf blister disease. In this disease small blister-like protrusions of the upper side of the leaf remain green but the corresponding concavities on the underside of the leaf are lined with bright yellow and orange spore masses. Infected leaves do not usually fall prematurely.

An eye-catching but uncommon and unimportant shoot blight caused by the fungus *Pollaccia* occurs from time to time, especially in wet springs. Spores infect the leaf blade causing black angular spots to develop. These enlarge and coalesce as the fungus spreads through the leaf down the petiole and into the young shoot. The shoot dies and the dead leaves and shoots hang conspicuously blackened and limp on the tree.

Serious cankering and dieback caused by the fungus *Dothichiza populea* is uncommon in British poplar cultivation. However, it can cause considerable damage on the Continent in nurseries or in plantations where growing conditions are poor and the plants therefore unthrifty. Another fungus *Cytospora chrysosperma* occurs very commonly on dead poplar wood. It is conspicuous because it exudes long orange thread like tendrils, which are in fact masses of spores. It has, however, only a very limited capacity to invade a healthy tree, and should not invariably be blamed for the death of the wood on which it is found.

Climatic damage to poplars is relatively rare. Compared with many other trees they can be regarded as resistant to injury from low temperatures. There are, however, some reports of 'frost-crack' in poplars on low-lying sites liable to winter and spring frosts. Injury, which takes the form of a longitudinal split in the bark and wood, is nearly always on the southern face of the stem. It seems most likely to occur when very cold nights are followed by warm sunny days.

Further details on diseases of poplar in Britain can be found in Phillips and Burdekin (1982), and a useful account of the management of poplar disease, with an emphasis on the use of resistant material has been prepared by Pinon (1984).

## Wind damage

The most common damage is wind breakage within or below the crown of fast-grown poplars. Although storm force winds can cause breakage during the winter months, damage more often occurs during summer gales. Fast growth may be a factor in increasing susceptibility particularly with the large-leaved cultivars of *P. trichocarpa*. Fortunately all poplars are capable of making a good recovery and this is particularly true of the *P. trichocarpa* cultivars. If the break occurs high enough up the tree to leave part of the original crown it is usually best to leave such trees standing.

In addition to wind breakage, trees may suffer local compression failure of a small section of the stem leading to the formation of wound calluses. These may be accompanied by infection by pathogens causing stain and eventually decay in that part of the stem.

Trees may be blown from the vertical by steady winds following planting and, depending on cultivar, trees may then continue to grow with a slight to moderate leaning stem. This may be overcome by ensuring that rooted and unrooted sets should be planted to a sufficient depth to remain stable during the first 2 years of growth.

## Squirrel damage

Poplars are generally susceptible to bark stripping activities by the grey squirrel (*Sciurus carolinensis*) during the period immediately prior to leaf flushing until early July. Damage is most serious in young, well established trees that may be partly or wholly girdled within the crown, leading to loss of crown from subsequent breakage. There is subjective evidence that some balsam poplars and their hybrids may be more susceptible than hybrid black poplars. The reason for such damage is not fully understood. Therefore, the only effective control is to eliminate the population of squirrels within maurauding range of poplar trees by the use of warfarin-baited hoppers during the critical period from March to July. Squirrel control at other times of the year will not reduce the risk of damage.

# REFERENCES

ANON. (1985). *Building near trees*. Practice Note 3. National House-Building Council, Amersham.

BEAN, W. J. (1976). *Trees and shrubs hardy in the British Isles*, 8th Edition, Vol. 3. Murray, London.

BEATON, A. (1987). Poplars and agroforestry. *Quarterly Journal of Forestry* 81(4), 225–233.

BEATON, A. (1988). A new look at poplars and their role in agroforestry. In, *Broadleaves — changing horizons*, ed. M. Potter, 4–8. Institute of Chartered Foresters, Edinburgh.

BROEKHUIZEN, J. T. M. van. (1964). The identification of poplar clones in the nursery. *Nederlands Bosbouw Tijdschrift* 36(4), 105–117.

BROEKHUIZEN, J. T. M. van. (1972). Morphological description and identification of a number of new commercial poplar clones. *Nederlands Bosbouw Tijdschrift* 44(7/8), 180–189.

CANNELL, M. G. R. (1980). Productivity of closely spaced young poplar on agricultural soils in Britain. *Forestry* 53(1), 1–21.

CHARDENON, J. and SEMIZOGLU, M. A. (1962). *Report of the study group on willow nomenclature*. 11th Session of the International Poplar Commission. Yugoslavia 1962. FAO/CIP/124.

CLAPHAM, Q. R., TUTIN, T. G. and WARBURG, E. F. (1962). *Flora of the British Isles*, 2nd Edition. Cambridge University Press, Cambridge.

COLLINGWOOD, G. H. and BRUSH, W. D. (1947). *Knowing your trees*. American Forestry Association.

CRICHTON, D. (1983). *The future of poplar in the United Kingdom*. MSc Thesis, Department of Forestry and Wood Science, University College of North Wales, Bangor.

CUTLER, D. F. and RICHARDSON, I. B. K. (1981). *Tree roots and buildings* (Kew tree root survey). Construction Press, London.

DAVIES, R. J. (1987a). *Black polythene mulches to aid tree establishment*. Arboriculture Research Note 71/87/ARB. DoE Arboricultural Advisory and Information Service, Forestry Commission.

DAVIES, R. J. (1987b). *Sheet mulches: suitable materials and how to use them*. Arboriculture Research Note 72/87/ARB. DoE Arboricultural Advisory and Information Service, Forestry Commission.

FAO (1979). *Poplars and willows in wood production and land use*. FAO Forestry Series No. 10. FAO, Rome.

FARMER, B. A. (1972). *Handbook of hardwoods*, 2nd Edition. HMSO, London.

GRAY, W. G. (1949). *The raising of aspen from seed*. Forestry Commission Forest Record 2. HMSO, London.

HAMILTON, G. J. and CHRISTIE, J. (1971). *Forest management tables (metric)*. Forestry Commission Booklet 34. HMSO, London.

JOBLING, J. (1960). *Establishment methods for poplars*. Forest Record 43. HMSO, London.

JOBLING, J. (1961). Recent developments in poplar planting. *Quarterly Journal of Forestry* 55(4), 287–293.

JOBLING, J. (1963/64). Poplars: varietal studies. *Report on Forest Research 1964*, 40–42. HMSO, London.

KOSTER, R. (1972). Eleven new poplar clones: a guide. *Populier* (Nederlands) 9(1), 2–7.

KRUSSMAN, H. (1986). *Manual of cultivated broadleaved trees and shrubs*. Vol. II, E-PRO, 425–437. Batsford, London.

LAVERS, G. M. (1969). *The strength properties of timber*. Forest Products Research Bulletin 50. HMSO, London.

MILLER, W. A. (1976). Fifty years of poplar. *Quarterly Journal of Forestry* 70(4), 201–206.

MILNE-REDHEAD, G. (1984). In pursuit of the poplar. *Natural World* 10, 26–28.

MILNE-REDHEAD, G. (1985/86). Our rarest native timber tree. *The Dendrologist* 3, 1 and 2.

MITCHELL, A. F. (1973). *Replacement of elm in the countryside*. Forestry Commission Leaflet 57. HMSO, London.

MITCHELL, A. F. (1974). *A field guide to the trees of Britain and northern Europe*. Collins, London.

MITCHELL, A. F. and HALLETT, V. E. (1985). *Champion trees in the British Isles*. Research and Development Paper 138. Forestry Commission, Edinburgh.

PEACE, T. R. (1962). *Pathology of trees and shrubs*. Clarendon Press, Oxford.

PHILLIPS, D. H. and BURDEKIN, D. A. (1982). *Diseases of forest and ornamental trees*, 281–306. The Macmillan Press Ltd., London and Basingstoke.

PINON, J. (1984). Management of diseases of poplar. *European Journal of Forest Pathology* **14**, 415–425.

PINON, J. (1986). Rust pathotypes of poplar. *EPPO Bulletin* **16**(3), 589–592.

PRIOR, R. (1983). *Trees and deer. How to cope with deer in forest, field and garden*. Batsford, London.

REHDER, A. (1962). *Manual of cultivated trees and shrubs hardy in North America*, 2nd revision, 10th printing. Macmillan, New York.

RIDÉ, M. (1958). Sur l'étiologie du chancre suintant du peuplier. *Comptes Rendus des Séances de l'Académie des Sciences* **246**, 2795–2798.

SARGENT, C. S. (1905). *Manual of the trees of North America*. Houghton Mifflin, New York.

SCHREINER, E. J. and STOUT, A. B. (1934). *Descriptions of ten new hybrid poplars*. Bulletin – Torrey Botanical Club 61, 449–460.

STEENACKERS, V. (1982). Nouvelle race physiologique de *Melampsora larici-populina* en Belgique. Proceedings of FAO International Poplar Commission at Casale Monferrato.

STERN, R. C. (1971). *Poplar growing at close spacing*. Home Grown Timber Advisory Committee Technical Sub-Committee Paper No. 311. (4pp.).

STERN, R. C. (1972). Poplar growing at close spacing. *Timber Grower* 44, 20–24.

STROUTS, R. G. (1980). Arboriculture — death in Ely. Why did so many Lombardy poplars die in the east of England last year? *Gardeners' Chronicle and Horticultural Trade Journal* **188**(7), 15 Aug., 53–56.

# Appendix 1

## The International Poplar Commission

The International Poplar Commission acts as the international registration authority for all poplar cultivars. In 1962, to ensure conformity of data presentation amongst member countries, the Commission adopted a form for recording the descriptions of both old and new black poplar hybrids. Subsequently, as increasing numbers of new cultivars were put forward for registration, including selections of *P. trichocarpa* and balsam black poplar hydrids, it was recognised by the Commission that additional clonal details were required, particularly to assist research workers and practitioners in other countries in the evaluation of new releases. The form was therefore expanded in 1975 to provide space for a summary account of the behaviour and uses of the cultivar as well as a full description to ensure reliable identification. Some 85 features can now be recorded on the form; 45 relating to identification, 21 to diseases, pests and non-biological hazards, 13 to wood utilisation and 6 to culture.

# Appendix 2

## New Poplar Clones from Belgium

In 1985, eleven new poplar clones, bred at the Government Poplar Research Station at Geraardsbergen, Belgium, were introduced into Britain. They are as listed below:

(*P.* × *euramericana*)

*P. deltoides* × *nigra* 'Gaver', 'Ghoy', 'Gibecq' and 'Primo';

*P. deltoides* × *trichocarpa* 'Beaupré', 'Boelare', 'Hunnegem', 'Raspalje' and 'Unal';

*P.* × *trichocarpa* 'Columbia River' and 'Trichobel'.

These eleven clones, together with two others, *P. deltoides* × *nigra* 'Isières' and 'Ogy', are collectively known as the UNAL clones as they were originally developed for the Union Alumetaire Match Company in Belgium.

By 1987 the eleven clones had been established in the populetum at Alice Holt and a new series of clonal trials was set up to examine their suitability for growing in this country. The first two of these trials (at Bedgebury Forest, Kent, and Ampthill Forest, Bedfordshire) were planted early in 1987 and the series is being extended to cover a wide range of site types throughout Britain.

Since their introduction into Britain, some of the new clones have been found to be susceptible to a new race of the leaf rust *Melampsora larici-populina*; 'Hunnegem', 'Raspalje' and 'Unal' are badly affected but the position of 'Trichobel' and 'Columbia River' is still unclear. The remaining six clones, which were free from infection, were provisionally approved by FC Research Division in 1989 and were added to the list of clones approved for commercial production under FRM regulations.

As they were bred for conditions in Belgium the UNAL clones are likely to grow best in the south of England and on good sites will probably attain similar growth rates to those in Belgium (up to twice that of 'Robusta'). In more northerly areas, there is likely to be a significant reduction in yield. Future introductions, especially *P. trichocarpa* hybrids which have been selected for British conditions, may have greater potential.

It is intended to introduce more clones from abroad for trials in this country.

# Index of Subjects and Technical Terms

This is a selective index and includes only the more important references. Characteristics of individual species and cultivars are not indexed here, but will be found described in Chapter 2, pages 4-32 (see also 'Index of Species and Cultivars').

amenity value 42
arable, effect on 40–41
arable intercropping 63–65
augers, soil 50

bacterial canker 71; see also Chapter 2
bark 34, 65
biomass 54, 55–56, 61–62
botanical
　classification 1–3
　key 33–36
branches 34–36
　epicormic 56
branchlets 34–36
buds 34, 36
buildings, situation near 41–42

canker; see also Chapter 2
　bacterial 71
　fungal 71–72
catkins 34–37, 49
characteristics
　species' 1–3; see also Chapter 2
　wood 66
ciliation 34
classification, botanical 1–3
climate 38–39, 72
clones 4–5
cordate 33
crown 34–36
cultivars 2; see also 'Index of Species and Cultivars'
cultivation, soil 52
cuneate 33
cuttings
　hardwood 44–46
　root 46–47
　semi-ripe 47–48
　softwood 47–48
　storage of 45, 48

damage: see also disease
　from insect pests 70–71
　from squirrels 73
　wind 72–73
disease 4, 5, 43, 71–72; see also Chapter 2
distribution, species' 1–3; see also Chapter 2
drainage 39–40, 63

elliptic 33
epicormic branches 56
explosives, use in planting 50
extraction 41

fertilisers 49, 52–53
flooding 40, 63
fodder 69
foliage, see leafing, leaves
fungi 71–72; see also 'Index of Species and Cultivars'
furniture 68, 69

glands 35
grant-aid 40, 46, 59
grazing 63, 65
growth rates 38–39, 42, 52, 54–57, 58–59, 60, 61
guards, plastic 65

height 58–59, 60, 61
herbicides 51
hybrids 4–5; see also 'Index of Species and Cultivars'

increment, see growth rates
insect pests 70–71; see also 'Index of Species and Cultivars'
intercropping, arable 63–65
International Poplar Commission 76

key, botanical 33–36

labelling 45–46
lanceolate 34
leaves, leafing 33–37
line planting 40–41, 63

match industry 63, 68
moisture
　in soil 38–40; see also drainage
　in wood 67
mulches 51–52, 53

nurseries 49

ovate 33

packaging, use in 68
pappus 49
particle board 69
peeling, rotary 58, 60, 68
pests, insect 70–71; see also 'Index of Species and Cultivars'
petioles 34–36
pH values 39, 49
planting 50–51
　explosives used in 50
　line 40–41, 63
plastic guards 65
plywood 68
production, propagation 44–49
pruning 38, 56–57, 64
pulpwood 54, 61, 69

rainfall 38–39; *see also* drainage
rhombic 33
roads, situation beside 41
root suckers 47
roots 40–42
rotary peeling 58, 60, 68
rotation 54, 56, 60

sawlogs 61
seed
  -capsules 34
  propagation from 48–49
sex 36–37
shading 41, 63
site 38–43, 58–59
situation 40–43
size standards 46

soil 39–40, 49
  augers 50
  cultivation 52
  pH values 39, 49
spacing 45, 48, 54–56, 59, 61, 64–65
species 1–32; *see also* 'Index of Species and Cultivars'
squirrel damage 73
standards, size 46
stem 35–36
stool beds 45–46
strength of wood 67
subcordate 33
suckers, root 47

temperature 38, 72
thinning 56
transplanting 45

truncate 33
twigs 35–37

uses of wood 67–69

veneer logs 59, 60, 61, 63, 68

weed control 51–52
wind damage 72–73
windbreaks 40, 42, 63
wood
  characteristics 66
  fibre, *see* biomass; pulpwood
  moisture content 67
  properties 67
  strength 67
  uses 67–69
woodland 42–43

yields 59–62, 63

# Index of Species and Cultivars

This index includes botanical and common names of trees and other species (including pests) mentioned in the text. Names of cultivars and varieties are qualified by, or cross-referenced to, their species or hybrid. The genus *Populus* is abbreviated to *P.* throughout.

abele, see *P. alba*
'Agathe F', *P.* × *euramericana* 19
*Agromyza* spp. (poplar cambium midge) 71
'Alcinde', *P. deltoides* 9
alder (*Alnus* spp.) 29, 39
*Alnus* spp. (alder) 29, 39
   *glutinosa* (common alder) 40
'AM', see *P.* × *euramericana* 'I-455'
American aspen, see *P. tremuloides*
'Andover', *P. trichocarpa* × *P. nigra* 30
'Androscoggin', *P. maximowiczii* × *P. trichocarpa* 22, 24, 25, 58, 60
*Anobium punctatum* (furniture beetle) 67
*Aplanobacter populi* 71
aspen, see *Trepidae*
   American, see *P. tremuloides*
   Chinese (*P. adenopoda*) 3, 7, 9
   European, see *P. tremula*
   Japanese (*P. sieboldii*) 3, 7
   large-toothed (*P. grandidentata*) 3, 5, 43
   quaking, see *P. tremuloides*
'Aurora', *P. candicans* 20

balm of Gilead, see *P. candicans*
balsam poplar, see *P. balsamifera*
   Japanese, see *P. maximowiczii*
   western, see *P. trichocarpa*
'Balsam Spire', see *P. balsamifera*
'Barn', *P.* × *interamericana* 28
'Beaupré', *P.* × *interamericana* 29, 76
beech, southern (*Nothofagus* spp.) 29

beetle
   furniture (*Anobium punctatum*) 67
   large longhorn (*Saperda populnea*) 71
   large poplar-leaf (*Chrysomela populi*) 70
   leaf 71
   powder post (*Lyctus brunneus*) 67
   small poplar-leaf (*Phyllodecta* spp.) 70
Berlin poplar (*P.* × *berolinensis*) 21, 26–27, 28, 36, 37, 70
*Betula pubescens* (hairy birch) 40
*betulifolia* (var.), see *P. nigra*
bilberry (*Vaccinium myrtillus*) 39
birch, hairy (*Betula pubescens*) 40
'BL Costanzo', *P.* × *euramericana* 19
black poplar, see *P. nigra*
   downy, see *P. nigra* var. *betulifolia*
   hybrids, see *P.* × *euramericana*
   Italian, see *P.* × *euramericana* 'Serotina'
'Blom', *P. trichocarpa* 24
'Boccalari', *P.* × *euramericana* 19
'Boelare', *P.* × *interamericana* 29, 76
'Bolleana', see *P. alba* 'Pyramidalis'
Bolle's poplar (*P. alba* 'Pyramidalis') 8, 34
'Branagesi', *P.* × *euramericana* 19
'Bruhl', *P. trichocarpa* 24
buckthorn (*Rhamnus catharticus*) 40

*Calluna* spp. (ling) 39
*Caltha palustris* (marsh marigold) 40
cambium midge, poplar (*Agromyza* sp.) 71
Canadian poplar, see *P.* × *euramericana* 'Marilandica'
*caudina* (var.), *P. nigra* 10
'Cappa Bigliona', *P.* × *euramericana* 19
*Carex* spp. (sedges) 40
'Carolin', *P. deltoides* 9
cedar, western red (*Thuja plicata*) 23
*Cerura vinula* (puss moth) 70
*Chamaecyparis lawsoniana* (Lawson cypress) 16
'Charkowiensis', *P. nigra* 11
'Chile', *P. nigra* 11
'Chileno', *P. nigra sempervirens* 11
Chinese aspen (*P. adenopoda*) 3, 7, 9
Chinese necklace poplar (*P. lasiocarpa*) 3, 31, 32
Chinese white poplar (*P. tomentosa*) 3, 8, 9
'Chrysocoma', *Salix* (golden weeping willow) 21
*Chrysomela populi* (large poplar-leaf beetle) 70
clearwing, hornet (*Sesia apiformis*) 71
'Columbia River', *P. trichocarpa* 24, 76
common alder (*Alnus glutinosa*) 40
'Cordata', *P. deltoides* 9, 28
*Cossus cossus* (goat moth) 71

cottonwood
  eastern, see *P. deltoides*
  *lance-leaf* (*P. acuminata*) 3, 19
  narrow-leaved (*P. angustifolia*) 3, 19–20
  swamp (*P. heterophylla*) 3, 31
'Criollo', *P. nigra* 11
*Cryptorhynchus lapathi* (weevil) 71
cypress, Lawson (*Chamaecyparis lawsoniana*) 16
*Cytospora chrysosperma* 12, 72

*davidiana* (var.), *P. tremula* 6
*Deschampsia flexuosa* (wavy hair-grass) 39
dock, great water (*Rumex hydrolapathum*) 40
'Donk', *P.* × *interamericana* 28
'Dorskamp', *P.* × *euramericana* 19
*Dothichiza populea* 72
Douglas fir (*Pseudotsuga menziesii*) 23
downy black poplar, see *P. nigra* var. *betulifolia*

eastern cottonwood, see *P. deltoides*
elm (*Ulmus* spp.) 41
'Erecta', *P. tremula* 6
*Erica* spp. (heather) 39
*Eucalyptus gunnii* (eucalyptus) 29
'Eugenei', *P.* × *euramericana* 13–14, 16, 36, 37, 51, 59
Euphrates poplar (*P. euphratica*) 3, 5
*Euphratodendron* 5
  *olivieri* 5
European aspen, see *P. tremula*

'Fastigiata', *P. simonii* 22
*Filipendula ulmaria* (meadow sweet) 40
fir, Douglas (*Pseudotsuga menziesii*) 23
'Flevo', *P.* × *euramericana* 19
'Florence Biondi', *P.* × *euramericana* 14, 19
'Fritzi Pauley', *P. trichocarpa* 23, 24, 29
'Frye', *P. laurifolia* × *P. nigra* 27
furniture beetle (*Anobium punctatum*) 67

'Gattoni', *P.* × *euramericana* 19
'Gaver', *P.* × *euramericana* 18, 76
'Gelrica', *P.* × *euramericana* 14, 35, 37, 51
'Geneva', *P. candicans* × *P.* × *berolinensis* 30
'Ghoy', *P.* × *euramericana* 18, 76
giant Lombardy poplar, see 'Gigantea'
'Gibecq', *P.* × *euramericana* 18, 76
'Gigantea', *P. nigra* 11, 12, 35, 37
Gilead, balm of, see *P. candicans*
*globosa* (var.), *P. tremula* 6
goat moth (*Cossus cossus*) 71
golden poplar (*P.* × *euramericana* 'Serotina Aurea') 18, 35
golden weeping willow (*Salix* 'Chrysocoma') 21
grass
  mat- (*Nardus stricta*) 39
  purple moor- (*Molinia caerulea*) 39
  wavy hair- (*Deschampsia flexuosa*) 39
  reed- (*Phalaris arundinacea*) 40
great water dock (*Rumex hydrolapathum*) 40
grey poplar, see *P.* × *canescens*
grey squirrel (*Sciurus carolinensis*) 73
*Gypsonoma aceriana* (poplar shoot borer) 70

hair-grass, wavy (*Deschampsia flexuosa*) 39
hairy birch (*Betula pubescens*) 40
'Hamoui', *P. nigra* 11
'Harvard', *P. deltoides* 10
*hastata* (var.), *P. trichocarpa* 20, 23–24, 25
heather (*Erica* spp.) 39
'Heidemij', *P.* × *euramericana* 14–15, 35, 37
'Heimburger', *P. trichocarpa* 24
hornet clearwing (*Sesia apiformis*) 71
'Hunnegem', *P.* × *interamericana* 76

'I-45/51', *P.* × *euramericana* 15–16
'I-78', *P.* × *euramericana* 15, 36, 37

'I-154', *P.* × *euramericana* 15–16
'I-214', *P.* × *euramericana* 15
'I-262', *P.* × *euramericana* 15–16
'I-455', *P.* × *euramericana* 15–16
*Iris pseudacorus* (wild iris, yellow flag) 40
'Isières', *P.* × *euramericana* 18, 76
Italian poplar, black, see *P.* × *euramericana* 'Serotina'
'Italica', see *P. nigra* 'Italica'

Japanese aspen (*P. sieboldii*) 3, 7
Japanese balsam poplar, see *P. maximowiczii*

'Laevigiata', see 'Heidemij'
lance-leaf cottonwood (*P. acuminata*) 3, 19
larch (*Larix* spp.) 72
large-toothed aspen (*P. grandidentata*), 3, 5, 43
*Larix* spp. (larch) 72
Lawson cypress (*Chamaecyparis lawsoniana*) 16
leopard moth (*Zeuzera pyrina*) 71
*Leucoma salicis* (white satin moth) 70
'Lincoln', *P. deltoides* 9
ling (*Calluna* spp.) 39
Lombardy poplar, see *P. nigra* 'Italica'
  female or giant, see *P. nigra* 'Gigantea'
longhorn beetle, large (*Saperda populnea*) 71
'Lux', *P. deltoides* 10
*Lyctus brunneus* (powder post beetle) 67

Manchester poplar, see *P. nigra* var. *betulifolia*
marigold, marsh (*Caltha palustris*) 40
'Marilandica', *P.* × *euramericana* 13, 16, 36, 37
'Marquette', *P. deltoides* 9
marsh marigold (*Caltha palustris*) 40
*Marssonina* spp. 19, 24, 26, 28, 30, 31
  *brunnea* 14, 15, 17, 18, 29, 71
  *populi-nigrae* 12, 14, 71

mat-grass (*Nardus stricta*) 39
'McKee poplar', *P. deltoides angulata* 58
meadow sweet (*Filipendula ulmaria*) 40
*Melampsora* spp. 15, 18, 20, 24, 26, 28, 30, 31, 71–72
　*allii-populina* 72
　*larici-populina* 17, 18, 29, 72, 76
*Michauxii* (var.), *P. balsamifera* 20, 25
midge, poplar cambium (*Agromyza* sp.) 71
*Molinia caerulea* (purple moor-grass) 39
moor-grass, purple (*Molinia caerulea*) 39
moth
　goat (*Cossus cossus*) 71
　leopard (*Zeuzera pyrina*) 71
　puss (*Cerura vinula*) 70
　white satin (*Leucoma salicis*) 70
'Muhle Larsen', *P. trichocarpa* 24

*Nardus stricta* (mat-grass) 39
narrow-leaved cottonwood (*P. angustifolia*) 3, 19–20
'NE 42', *P. maximowiczii* × *P. trichocarpa* 25
'NE 226', *see* 'Florence Biondi'
'NE 388' *P. maximowiczii* × *P. trichocarpa* 25
necklace poplar, Chinese (*P. lasiocarpa*) 3, 31, 32
*neapolitana* (var.), *P. nigra* 10
Norway spruce (*Picea abies*) 66, 67
*Nothofagus* spp. (southern beech) 29

oak (*Quercus* spp.) 41
× *octorasdos*, *P. laurifolia* × *P. nigra* 28
'Ogy', *P.* × *euramericana* 18, 76
'Onda', *P. deltoides* 10
Ontario poplar (*P. ontariensis*) 20
'OP 226', *see* 'Florence Biondi'
*Ophiobolus graminis* ('take-all') 64
'Oxford', *P. candicans* × *P.* × *berolinensis* 30–31

'Parasol de St Julian', *P. tremuloides* 6

*Pemphigus bursarius* 70
*Pemphigus spyrothecae* 70
*pendula* (var.), *P. tremula* 6
'Pendula', *P. simonii* 22
'Pendula', *P. tremula* 6
'Peoria', *P. deltoides* 9
× *petrowskyana*, *P. laurifolia* × *P. nigra* 27
*Phalaris arundinacea* (reed-grass) 40
*Phragmites communis* (reed) 40
*Phyllodecta vitellinae* 70
*Phyllodecta vulgatissima* 70
*Picea abies* (Norway spruce) 66, 67
*Picea sitchensis* (Sitka spruce) 23
pine, Scots (*Pinus sylvestris*) 6, 67
*Pinus sylvestris* (Scots pine) 6, 67
'Plantierensis', *P. nigra*, 12, 16, 28, 35
*Pollaccia* sp. 72
poplar
　balsam, *see P. balsamifera*
　Berlin, *see P.* × *berolinensis*
　black, *see P. nigra*
　black, hybrid, *see P.* × *euramericana*
　black Italian, *see P.* × *euramericana* 'Serotina'
　Bolle's (*P. alba* 'Pyramidalis') 8, 34
　Canadian, *see P.* × *euramericana* 'Marilandica'
　Chinese necklace (*P. lasiocarpa*) 3, 31, 32
　Chinese white (*P. tomentosa*) 3, 8, 9
　downy black, *see P. nigra* var. *betulifolia*
　Euphrates (*P. euphratica*) 3, 5
　golden (*P.* × *euramericana* 'Serotina Aurea') 18, 35
　grey, *see P.* × *canescens*
　Italian, black, *see P.* × *euramericana* 'Serotina'
　Japanese balsam, *see P. maximowiczii*
　Lombardy, *see P. nigra* 'Gigantea'; 'Italica'
　Manchester, *see P. nigra* var. *betulifolia*
　Ontario (*P. ontariensis*) 20

railway (*P.* × *euramericana* 'Regenerata' 16, 35, 37, 39
Swiss, *see P.* × *euramericana* 'Serotina'
volunteer, *see P. laurifolia*
white, *see P. alba*
willow-leaved (*P. angustifolia*) 3, 19–20
poplar cambium midge (*Agromyza* sp.) 71
poplar-leaf beetle
　large (*Chrysomela populi*) 70
　small (*Phyllodecta* spp.) 70
poplar shoot borer (*Gypsonoma aceriana*) 70
*POPULUS* spp. (POPLAR)
　*acuminata* (lance-leaf cottonwood) 3, 19
　*adenopoda* (Chinese aspen) 3, 7, 9
　*Aigeiros* (black poplars) 2, 3, 9–19, 26-31, 34, 39, 44, 45, 46, 47, 61, 66
　*alba* (white poplar, abele) 3, 7–9, 34, 37, 69
　　'Bolleana', *see P. alba* 'Pyramidalis'
　　'Pyramidalis' (Bolle's poplar) 8, 34
　　'Raket' 8
　　'Richardii' 8
　*alba* × *P. alba* 'Pyramidalis', *see P. alba* 'Raket'
　*alba* × *P. tremula* var. *davidiana*, *see P. tomentosa*
　*alba* × *P. trichocarpa* 26
　*Albidae* (white poplars) 2, 3, 7–9, 34, 59, 61, 66
　*angustifolia* (narrow-leaved cottonwood, willow-leaved poplar) 3, 19–20
　*balsamifera* (balsam poplar) 3, 20, 23, 34, 36, 37, 39, 43
　　'Balsam Spire' 20, 24, 25–26, 36, 37, 61, 63, 65, 70
　　'tacatricho 32', *see P. balsamifera* 'Balsam Spire'
　　'TT 32', *see P. balsamifera* 'Balsam Spire'
　　var. *Michauxii* 20, 25
　× *berolinensis* (Berlin poplar) 21, 26–27, 28, 36, 37, 70

81

*bolleana*, see *P. alba* 'Pyramidalis'
*canadensis*, see *P.* × *euramericana*
*candicans* (balm of Gilead) 20, 36, 37, 39; see also *P. maximowiczii*
  'Aurora' 20
*candicans* × *P.* × *berolinensis* 30–31
  'Geneva' 30
  'Oxford' 30–31
× *canescens* (grey poplar) 2, 7, 8–9, 34, 37, 39, 41, 46, 47, 48, 66, 67
*cathayana* 3, 20–21, 22
*certinensis*, see *P.* × *berolinensis*
*ciliata* 3, 21
*deltoides* (eastern cottonwood) 3, 9–10, 12, 13, 20, 24, 28, 29, 31, 35, 44, 45
  'Alcinde' 9
  'Carolin' 9
  'Cordata' 9, 28
  'Harvard' 10
  'Lincoln' 9
  'Lux' 10
  'Marquette' 9
  'Onda' 10
  'Peoria' 9
*deltoides* × *P. nigra*, see *P.* × *euramericana*
*deltoides angulata* 16, 58
  'McKee poplar' 58
*deltoides missouriensis*, see *P.* × *euramericana* 'Heidemij'
*euphratica* (Euphrates poplar) 3, 5
× *euramericana* (black poplar hybrids) 9, 12–19, 23, 28, 36, 38–39, 51, 54, 56, 59, 60, 64, 65, 66, 73, 76
  'Agathe F' 19
  'AM', see *P.* × *euramericana* 'I-455'
  'BL Costanzo' 19
  'Boccalari' 19
  'Branagesi' 19
  'Cappa Bigliona' 19
  'Dorskamp' 19
  'Eugenei' 13–14, 16, 36, 37, 51, 59
  'Flevo' 19
  'Florence Biondi' 14, 19
  'Gattoni' 19
  'Gaver' 18, 76
  'Gelrica' 14, 35, 37, 51
  'Ghoy' 18, 76
  'Gibecq' 18, 76
  'Heidemij' 14–15, 35, 37
  'I-45/51' 15–16
  'I-78' 15, 36, 37
  'I-154' 15–16
  'I-214' 15
  'I-262' 15–16
  'I-455' 15–16
  'Isières' 18, 76
  'Laevigiata', see *P.* × *euramericana* 'Heidemij'
  'Marilandica' (Canadian poplar) 13, 16, 36, 37
  'NE 226', see *P.* × *euramericana* 'Florence Biondi'
  'Ogy' 18, 76
  'OP 226', see *P.* × *euramericana* 'Florence Biondi'
  'Primo' 18, 76
  'Regenerata' (railway poplar) 16, 35, 37, 39
  'Robusta' 14, 15, 16–17, 18, 24, 29, 34, 37, 60, 64, 76
  'San Martino' 19
  'Serotina' (Swiss, Canadian, or black Italian poplar) 13, 14, 16, 17–18, 35, 37, 39, 51, 59, 67
  'Serotina Aurea' (golden poplar) 18, 35
  'Serotina de Selys' 18, 35
  'Spijk' 19
  'Triplo' 19
*fremontii* 3, 10
× *generosa* 24, 28, 36, 37
*glandulosa* × *P. alba* 9
*grandidentata* (large-toothed aspen) 3, 5, 43
*grandidentata* × *P. alba* 5, 8
*grandidentata* × *P. tremula* 7
*heterophylla* (swamp cottonwood) 3, 31
× *interamericana* 28–30, 60, 61
  'Barn' 28
  'Beaupré' 29, 76
  'Boelare' 29, 76
  'Donk' 28
  'Hunnegem' 76
  'Rap' 28, 29, 55, 62
  'Raspalje' 76
  'Unal' 76
*koreana* 3, 21, 37
*lasiocarpa* (Chinese necklace poplar) 3, 31, 32
*laurifolia* 3, 21, 26, 27, 28
*laurifolia* × *P. nigra* 26–28
  × *berolinensis* (Berlin poplar) 21, 26–27, 28, 36, 37, 70
  'Frye' 27
  × *octorasdos* 28
  × *petrowskyana* 27
  × *rasumowskyana* 27–28
  'Rumford' 27
  'Strathglass' 27
  × *wobstii* 28
*Leuce* 2, 3, 5–9, 26, 44, 45, 46, 47, 48
*Leucoides* 2, 3, 31–32, 44, 59
*maximowiczii* (Japanese balsam poplar) 3, 21–22, 25, 28, 30, 31, 36, 37
*maximowiczii* × *P. nigra* 28
  'Rochester' 28
*maximowiczii* × *P. trichocarpa* 25
  'Androscoggin' 22, 24, 25, 58, 60
  'NE 42' 25
  'NE 388' 25
*nigra* (black poplar) 2, 3, 10–12, 27, 35, 37, 42, 43, 49, 69, 70
  'Charkowiensis' 11
  'Chile' 11
  'Criollo' 11
  'Gigantea' (female or giant Lombardy poplar) 11, 12, 35, 37
  'Hamoui' 11
  'Italica' (Lombardy poplar) 10, 11–12, 13, 18, 26, 27, 28, 34, 35, 37, 40, 71
  'Plantierensis' 12, 16, 28, 35

var. *betulifolia* (downy black, or Manchester poplar) 10, 12, 30, 34, 35, 37
var. *betulifolia* × *P. nigra* 'Italica', see *P. nigra* 'Plantierensis'
var. *caudina* 10
var. *neapolitana* 10
var. *thevestina* 11
'Vereecken' 12, 37
*nigra* × *P. deltoides*, see *P.* × *euramericana*
*nigra sempervirens* 11
'Chileno' 11
*obtusata fastigiata*, see *P. simonii* 'Fastigiata'
*ontariensis* (Ontario poplar) 20
*purdomii* 3, 22
*sargentii* 3, 10
*sieboldii* (Japanese aspen) 3, 7
*simonii* 3, 22
    'Fastigiata' 22
    'Pendula' 22
*simonii obtusata*, see *P. simonii* 'Fastigiata'
*suaveolens* 3, 22–23
    'Pyramidalis' 23
*szechuanica* 3, 23
*Tacamahaca* (balsam poplars) 2, 3, 19–31, 44, 45, 46, 47, 59, 60, 66, 73
*tacamahaca*, see *P. balsamifera*
*tacamahaca* × *P. trichocarpa* 32, see *P. balsamifera* 'Balsam Spire'
*tomentosa* (Chinese white poplar) 3, 8, 9
*tremula* (European aspen) 2, 3, 5, 6, 7, 9, 34, 37, 39, 43, 46, 47, 48, 49
    'Erecta' 6
    'Pendula' 6
    'Purpurea' 6
    var. *davidiana* 6
    var. *globosa* 6
    var. *pendula* 6
    var. *villosa* 6
*tremula* × *P. alba*, see *P.* × *canescens*
*tremula* × *P. tremuloides* 7, 48

*tremuloides* (quaking, or American aspen) 3, 6–7, 43
    'Parasol de St Julien' 6
*Trepidae* (aspens) 5–9, 59, 66, 70
*trichocarpa* (western balsam poplar, black cottonwood) 3, 10, 19, 21, 23–24, 25, 26, 28, 29, 30, 36, 37, 39, 42, 43, 54, 56, 58, 59, 60, 61, 64, 65, 72, 76
    'Blom' 24
    'Bruhl' 24
    'Columbia River' 24, 76
    'Fritzi Pauley' 23, 24, 29
    'Heimburger' 24
    'Muhle Larsen' 24
    'Scott Pauley' 23, 24
    'Trichobel' 24, 76
    var. *hastata* 20, 23–24, 25
*trichocarpa* × *P. deltoides*, see *P.* × *interamericana*
*trichocarpa* × *P. nigra* 30
    'Andover' 30
    'Roxbury' 30
*tristis* 3, 24
*Turanga* 2, 3, 5
*violascens* 3, 31–32
*wilsonii* 3, 32
*wislizeni* 3, 10
*yunnanensis* 3, 25
powder post beetle (*Lyctus brunneus*) 67
'Primo', *P.* × *euramericana* 18, 76
*Pseudotsuga menziesii* (Douglas fir) 23
purple moor-grass (*Molinia caerulea*) 39
'Purpurea', *P. tremula* 6
puss moth (*Cerura vinula*) 70
'Pyramidalis', *P. alba* 8, 34
'Pyramidalis', *P. suaveolens* 23

quaking aspen, see *P. tremuloides*
*Quercus* spp. (oak) 41

railway poplar, see 'Regenerata'
'Raket', *P. alba* 8
'Rap', *P.* × *interamericana* 28, 29, 55, 62
'Raspalje', *P.* × *interamericana* 76
× *rasumowskyana*, *P. laurifolia* × *P. nigra* 27–28

reed (*Phragmites communis*) 40
reed-grass (*Phalaris arundinacea*) 40
'Regenerata', *P.* × *euramericana* 16, 35, 37, 39
*Rhamnus catharticus* (buckthorn) 40
*Rhododendron* spp. 39
'Richardii', *P. alba* 8
'Robusta', see *P.* × *euramericana*
'Rochester', *P. maximowiczii* × *P. nigra* 28
'Roxbury', *P. trichocarpa* × *P. nigra* 30
*Rumex hydrolapathum* (great water dock) 40
'Rumford', *P. laurifolia* × *P. nigra* 27

*Salicaceae* 1, 5
*Salix* spp. (willow) 1, 29, 39, 40, 41
    'Chrysocoma' (golden weeping willow) 21
'San Martino', *P.* × *euramericana* 19
*Saperda populnea* (large longhorn beetle) 71
satin moth, white (*Leucoma salicis*) 70
*Sciurus carolinensis* (grey squirrel) 73
Scots pine (*Pinus sylvestris*) 6, 67
'Scott Pauley', *P. trichocarpa* 23, 24
sedges (*Carex* spp.) 40
'Serotina', see *P.* × *euramericana* 'Serotina'
'Serotina Aurea', *P.* × *euramericana* 18, 35
'Serotina de Selys', *P.* × *euramericana* 18, 35
*Sesia apiformis* (hornet clearwing) 71
shoot borer, poplar (*Gypsonoma aceriana*) 70
Sitka spruce (*Picea sitchensis*) 23
Southern beech (*Nothofagus* spp.) 29
'Spijk', *P.* × *euramericana* 19
spruce
    Norway (*Picea abies*) 66, 67
    Sitka (*Picea sitchensis*) 23

squirrel, grey (*Sciurus carolinensis*) 73
'Strathglass', *P. laurifolia* × *P. nigra* 27
swamp cottonwood (*P. heterophylla*) 3, 31
Swiss poplar, *see P.* × *euramericana* 'Serotina'

'tacatricho 32', *see* 'Balsam Spire'
'take-all' (*Ophiobolus graminis*) 64
*thevestina* (var.), *P. nigra* 11
*Thuja plicata* (western red cedar) 23
'Trichobel', *P. trichocarpa* 24, 76
'Triplo', *P.* × *euramericana* 19
'TT 32', *see* 'Balsam Spire'

*Ulmus* spp. (elm) 41

'Unal', *P.* × *interamericana* 76

*Vaccinium myrtillus* (bilberry) 39
'Vereecken' *P. nigra*, 12, 37
*villosa* (var.), *P. tremula* 6
volunteer poplar, *see P. laurifolia*

water dock, great (*Rumex hydrolapathum*) 40
wavy hair-grass (*Deschampsia flexuosa*) 39
weevil (*Cryptorhynchus lapathi*) 71
western balsam poplar, *see P. trichocarpa*
western red cedar (*Thuja plicata*) 23
white poplar, *see P. alba*
  Chinese (*P. tomentosa*) 3, 8, 9

white satin moth (*Leucoma salicis*) 70
wild iris (*Iris pseudacorus*) 40
willow (*Salix* spp.) 1, 29, 39, 41
  golden weeping (*Salix* 'Chrysocoma') 21
willow-leaved poplar (*P. angustifolia*) 3, 19–20
× *wobstii*, *P. laurifolia* × *P. nigra* 28

*Xanthomonas* (*Aplanobacter*) *populi* 71

yellow flag (*Iris pseudacorus*) 40

*Zeuzera pyrina* (leopard moth) 71